THE CLIMATE-CONSCIOUS CORPORATE

PURBA RAO

ARJUN ISAAC CHAN-HALADY

notionpress.com

INDIA · SINGAPORE · MALAYSIA

ISBN
Paperback 979-8-90392-264-2
Hardcase 979-8-90510-382-7

Dedication

This book is dedicated to our

Brilliant, amazing and wonderful Aria

Bangalore 2026

CONTENTS

PREFACE

Mission for the book.

Let me start by recounting an incident which happened quite a few years back.

In those days, I was teaching elective courses in various Indian Business Schools. One of these was a premium Business School near the coastline, in the southern part of India. It was a beautiful Resort Campus loved by the students. In my first year at this school, the classes had started, but the campus was still being constructed.

In this first year of my teaching, it was November. Every year, in November/December, it used to rain like a cloudburst in that place, and the campus would get flooded. Also, the campus was at a lower level than the highway running across. So, all the water from the highway came down to the campus and stayed there.

In that first year of my teaching there, the walkways from my room where I stayed were all flooded, and there was water everywhere. No electricity on the campus, classes suspended, and darkness everywhere. Somehow, I waded through the water and managed to go to the canteen where many of the students were gathered. It was dark there, and I could barely see the students who were talking to each other and also to me.

They related similar incidents in their hometowns where they encountered flash floods, cloud bursts, rivers rising, buildings collapsing and so on. The students said they would be eager to contribute towards preventing such calamities and help the communities which suffered. Trapped in their home, no food, no medical supplies, just waiting for relief to arrive. The students said that as future leaders and managers, they would be very eager to ensure that such events do not happen again and again.

All the same, they said that there was not much awareness regarding the impacts of climate change, or even on the factors leading to climate change. The subject was not taught in the Business Schools. However, they emphasised that they would be able to help communities suffering on account of climate change, with food, water, medicines, logistics, etc. and would look forward to doing so.

Listening to them, I decided that I would conduct research and publish articles to help build awareness among professional managers in India... awareness relating to climate change... awareness of the factors leading to climate change, the devastating impacts and what corporate managers could do to address them.

This has been the mission for writing the book.

Here, in this book, we have included 10 chapters:

Chapters 1 and 2 talk about basic concepts of climate change... the factors leading to climate change and the key sectors where the impacts are felt.

Chapter 3 deals with Strategies to deal with climate change... Mitigation and Adaptation.

Chapter 4 discusses how corporate managers/private sector can help address climate change concerns.

Chapters 5-8 talk about the impacts of climate change on the Water sector, Ecosystem sector, Food and Agriculture sector and Disease and Health sector.

Chapter 9 includes six case studies on how the corporate sector is addressing climate change in some key manufacturing and service sectors: Semiconductor sector, Textiles, FMCG sector, Electric Vehicles sector and Hospitality sector. The last case explores the initiatives in a resort hotel addressing climate change.

Chapter 10 has been written by my co-author, **Arjun Isaac Chan-Halady,** a high school student exploring artificial intelligence in climate change. He has discussed how AI can help in addressing Climate Change.

Purba Rao, Faculty, Myra School of Business, Mysuru, Karnataka, India

March 2026.

CERTAIN TERMS USED IN THE MANUSCRIPT

ADB: Asian Development Bank

CDP: Carbon Disclosure Project is a global non-profit operating the premier environmental disclosure system for companies, cities, and regions to manage their climate risks, emissions, and water/forest impacts.

CERES: Clouds And Earth's Radiant Energy System

CER: Certified Emissions Reduction

CF: Carbon footprint

CH4: Methane

CII: Confederation of Indian Industry

CE: Circular Economy: In a circular economy, products and materials are kept in circulation through processes like maintenance, reuse, refurbishment, remanufacture, recycling, and composting. The circular economy tackles climate change and other global challenges, like biodiversity loss, waste, and pollution.

CO_2: Carbon Dioxide

CRA: Climate-Resilient Agriculture (CRA)

CSR: Corporate Social Responsibility.

CSRD: Corporate Sustainability Reporting Directive

EPA: Environmental Protection Agency

EU: European Union

FAO: The Food and Agriculture Organization , United Nations

FMCG: Fast-Moving Consumer Goods

GHG: Greenhouse Gas

GTZ (Deutsche Gesellschaft für Technische Zusammenarbeit or German Corporation for Technical Cooperation) is the agency which channels international aid from the Federal German Government to developing countries.

GWP: Global Warming Potential

HVAC: Heating, Ventilation, and Air Conditioning

IPCC: Intergovernmental Panel on Climate Change

LED: Light-Emitting Diode

NGO: Non-Government Organization

NOAA: National Oceanic and Atmospheric Administration

PCAF: Partnership for Carbon Accounting Financials (PCAF)

SDG Report (Sustainable Development Goals, United Nations)

SMB: Small and Medium-Sized Business

SME: Small and Medium Enterprises

WB: World Bank

WMO: World Meteorological Organization

UNICEF: United Nations Children's Educational Fund

CHAPTER 1

INTRODUCTION: EXPLAINING GLOBAL WARMING AND CLIMATE CHANGE

Global warming is the phenomenon of increasing average air temperatures near the surface of Earth, which has been happening over the past centuries.

It is the slow increase in the average temperature of the Earth's lower atmosphere because an increased amount of the energy (heat) striking the Earth from the sun is being trapped in the atmosphere and not radiated out into space. The Earth's atmosphere, over time, has accumulated various gases which act like a greenhouse to prevent the sun's heat from escaping into the outer atmosphere, leading to Earth having an increase in temperature. In the past, the accumulation of gases was not that huge or significant. In fact, this increase in temperature permitted the emergence of life forms as we know them, including human beings living on the face of the Earth. However, now the increase in gases is too huge, causing an unacceptable increase in temperature, which raises various concerns.

(https://www.britannica.com/science/global-warming)

Since the mid-20th century, climate scientists have gathered detailed observations of various weather conditions (such as temperatures, precipitation, and floods and storms) and of related influences on climate (such as rainfalls, extreme heat conditions, ocean currents and the atmosphere's chemical composition). These data indicate that Earth's climate has changed over almost every conceivable timescale since the beginning of geologic time. Also, this has been happening since the beginning of the Industrial Revolution, because of human activities which have had a growing influence over the pace and extent of present-day climate change.

Thus, global warming and climate change are interrelated, with global warming being the cause leading to the climate change phenomenon.

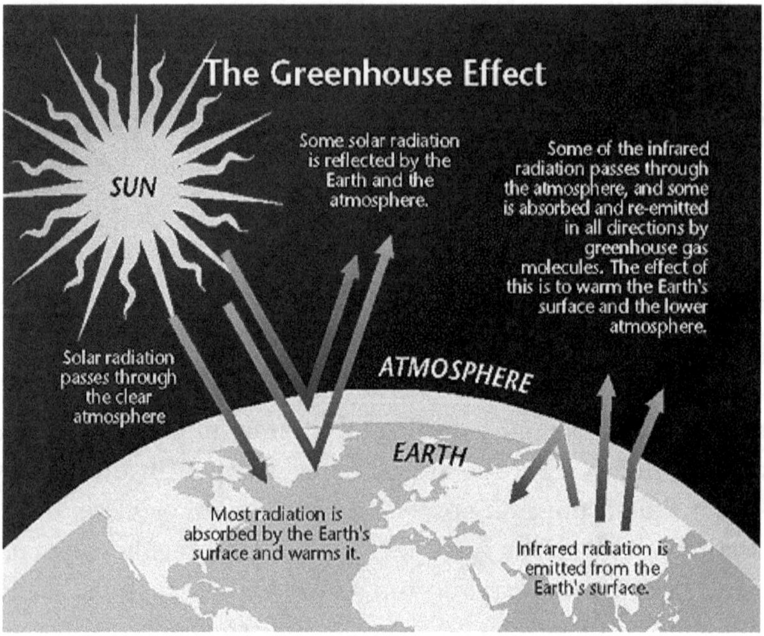

Figure 1.1 The Greenhouse Effect

1.1 Why Did the Accumulated Gases Increase?

The significant gases which act as a shield to prevent some of the heat from the sun from escaping into the atmosphere are called Green House Gases, GHGs. These gases, along with water vapour (H2O), keep some of the heat from the sun from escaping into the atmosphere.... Like the glass walls of a greenhouse. The more the GHGs in the atmosphere, the more heat gets trapped, strengthening the greenhouse effect and increasing the temperature of the Earth. This leads to global warming phenomena, which further leads to climate change, comprising long-term shifts in temperature and weather patterns.

Although some changes in Earth's climate are natural, like volcanic eruptions, most causes of climate change are anthropogenic—or caused by humans.

"Scientists agree that human activity is now the leading cause of climate change and is why the planet is warming up faster than it ever has in the past."

(https://science.nasa.gov/climate-change/causes/)

1.2 The most common and most talked about GHGs are:

*CO_2 or carbon dioxide, which is produced any time something is burned.

In fact, because it is so common, it is used as a measure of gases that warm the atmosphere. CO_2 or carbon dioxide is the most common GHG, constituting, by some measures, almost 55% of total long-term GHGs. It is used as a standard marker by the United States Environmental Protection Agency, and is assigned a GWP or Global Warming Potential of 1.

GWP, global warming potential, is a metric comparing the heat-trapping capability of greenhouse gases to carbon dioxide over a specific timeframe (usually 100 years). Carbon dioxide is assigned a GWP of 1. Higher GWP values indicate greater, more intense warming contributions.

* Methane (CH4), another important GHG, is 28-36 times as warming as CO_2 when in the upper atmosphere. This is according to the United States Environmental Protection Agency, which estimates the GWP, Global Warming Potential, of methane, as estimated over 100 years, as 28-36. Therefore, 1 ton of methane = 28-36 tons eCO_2 or CO_2 equivalents. Thus, CH4 has a GWP of 28-36.

Methane is produced in many combustion processes and also by anaerobic decomposition, for example, in flooded rice paddies, pig and cow stomachs, and pig manure ponds. Methane breaks down in approximately 10 years, but is a precursor of ozone, itself an important GHG.

- Nitrous oxide, or Paren (laughing gas), NO/N2O or simply NOx, is a byproduct of fertiliser production and use, other industrial processes and the combustion of certain materials. Nitrous oxide lasts a very long time in the atmosphere, but at the 100-year point of comparison to CO_2, its GWP is 265-298.

- Fluorinated gases were created as replacements for ozone-depleting refrigerants, but have proved to be both extremely long-lasting and extremely warming GHGs. They have no natural sources, but are entirely manmade. At the 100-year point of comparison, their GWPs range from 1,800 to 8,000 and some variants top 10,000. Fluorinated gases (F-gases) is a term used to refer to fluorinated greenhouse gases. Major classes include Hydrofluorocarbons (HFCs), Perfluorocarbons (PFCs), and sulphur.
- Sulphur hexafluoride or SF6 is used for specialised medical purposes, but primarily in what are called dielectric materials, especially dielectric liquids. This is a colourless, odourless, non-flammable, and non-toxic gas with high global warming potential. It is used as an insulating gas in the electric power industry and as a speciality gas in other applications.

These are also used as insulators in high voltage applications such as transformers and grid switching gear. SF6 will last thousands of years in the upper atmosphere and has a GWP of 22,800.

As indicated above, associated with global warming is climate change, which refers to significant, long-term changes in the global climate. The global climate is the connected system of sun, earth and oceans, wind, rain and snow, forests, deserts and savannas, and everything associated with life.

1.3 How Does Global Warming Drive Climate Change?

Heat is energy, and when energy is added to any system, changes occur. Because all systems in the global climate system are connected, adding heat energy causes the global climate as a whole to change. For instance, consider that much of the world is covered with oceans, which heat up with the added heat energy. When the ocean heats up, more water evaporates into clouds, where storms like hurricanes and typhoons form, leading to much more unpredictable and intense storms.

The climate of a place can be described as its rainfall, changing temperatures during the year and so on. But the global climate is more than the "average" of the climates of specific places. A description of the global climate includes how, for example, the rising temperature of the oceans feeds typhoons which blow harder, drop more rain and cause more damage, and also shifts global ocean currents that melt Antarctica's ice, which slowly makes the sea level rise. It is this systemic connectedness that makes global climate change so important and so complicated.

A warmer atmosphere makes glaciers and mountain snow packs, the polar ice cap, and the great ice shield jutting off of Antarctica melt, raising sea levels. Changes in temperature change the great patterns of wind that bring the monsoons in Asia and rain and snow around the world, making drought and unpredictable weather more common.

(Causes of climate change, https://www.epa.gov/climatechange-science/basics-climate-change)

1.4 What generates GHGs which lead to climate change

***Generating power**

Burning coal, oil and gas, fossil fuels, etc., to generate electricity produces GHGs. In particular, generating electricity produces carbon dioxide and nitrous oxide. Generating electricity and heat by burning fossil fuels causes a large amount of global emissions. Most electricity is still generated by burning coal, oil, or gas, which produces carbon dioxide and nitrous oxide - powerful greenhouse gases that blanket the Earth and trap the sun's heat.

Incidentally, as opposed to this use of fossil fuels, roughly one-seventh of the world's total primary energy now comes from renewable sources, including solar, wind, hydropower, and modern biofuels. That works out to about 14% of all energy consumed globally. (https://scienceinsights.org/how-much-energy-comes-from-renewable-sources/)

***Manufacturing goods**

We know burning coal, oil and gas produces carbon dioxide and nitrous oxide. This is what happens when manufacturing takes place. Manufacturing and industrial activities, in general, produce emissions, mostly from burning fossil fuels to produce energy for making things like cement, iron, steel, electronics, plastics, clothes, and other goods. Mining and other industrial processes also release gases, as does the construction industry. Machines used in the manufacturing process often run on coal, oil, or gas, and some materials, like plastics, are made from chemicals sourced from fossil fuels.

The manufacturing industry is one of the largest contributors to greenhouse gas emissions worldwide.

***Cutting down forests (deforestation).**

Trees help to regulate the climate by absorbing CO_2 from the atmosphere. When they are cut down, that beneficial effect is lost and the carbon stored in the trees is released into the atmosphere, adding to the greenhouse effect. That is why loss of trees leads to the absence of CO_2-absorbing sinks, so essential to reducing the volume of CO_2 in the atmosphere.

Cutting down forests to create farms or pastures, or for any other reasons, causes emissions, since trees, when they are cut, release the carbon they have been storing. Each year, approximately 12 million hectares of forest are destroyed. Since forests absorb carbon dioxide, destroying them also limits nature's ability to keep emissions out of the atmosphere. Deforestation, together with agriculture and other land use changes, is responsible for roughly a quarter of global greenhouse gas emissions.

***Using transportation**

Most cars, trucks, ships, and planes run on fossil fuels. That makes transportation a major contributor to greenhouse gases, especially carbon dioxide emissions. Road vehicles account for the largest part, due to the combustion of petroleum-based products, like gasoline, in internal combustion engines. But emissions from ships and planes continue to grow. Transport accounts for nearly one-quarter of global energy-related carbon dioxide emissions. The trends point to a significant increase in energy use for transport over the coming years.

***Producing food**

Producing food causes emissions of carbon dioxide, methane, and other greenhouse gases in various ways, through deforestation and clearing of land for agriculture and grazing, digestion by cows and sheep, the production and use of fertilisers and manure for growing crops, and the use of energy to run farm equipment or fishing boats, usually with fossil fuels. All this makes food production a major contributor to climate change. Greenhouse gas emissions also come from packaging and distributing food.

*Raising livestock farming

Cows and sheep produce large amounts of methane when they digest their food.

***Using Fertilisers** containing nitrogen produces nitrous oxide emissions.

***Waste management**, such as landfills and trash incinerators, gives out greenhouse gases.

***Clearing land for agriculture**. Contributes significantly to climate change through the release of carbon dioxide, methane, and nitrous oxide.

***Fluorinated gases are** also emitted from equipment and products that use these gases. Such emissions have a very strong warming effect, up to 23,000 times greater than CO_2.

***Powering buildings**

Globally, residential and commercial buildings consume over half of all electricity. As they continue to draw on coal, oil, and natural gas for heating and cooling, they emit significant quantities of greenhouse gases. Growing energy demand for heating and cooling, with rising air-conditioner ownership, as well as increased electricity consumption for lighting, appliances, and connected devices, has contributed to a rise in energy-related carbon dioxide emissions from buildings in recent years.

***Consuming too much**

Our home and use of power, how we move around, what we eat and how much we throw away ... all contribute to greenhouse gas emissions. So does the consumption of goods such as clothing, electronics, and plastics. A large chunk of global greenhouse gas emissions is linked to private households. Our lifestyles have a profound impact on our planet. The wealthiest bear the greatest responsibility: the affluent people, roughly a small percentage of the global population combined, account for more greenhouse gas emissions than the poorest.

1.5 What Impacts Does Climate Change Have?

Because the global climate is a connected system, the impacts of climate change are felt everywhere.

Specific impacts of Climate Change are

- **Rising sea levels:** Average sea level around the world rose about 8 inches (20 cm) in the past 100 years; climate scientists expect it to rise more and more rapidly in the next 100 years. Coastal cities such as New York are already seeing an increased number of flooding events, and by 2050, many such cities may require sea walls to survive. Estimates vary, but conservatively, sea levels are expected to rise 1 to 4 feet (30 to 100 cm), enough to flood many small Pacific island states (Vanuatu), famous beach resorts and coastal cities (Bangkok, Boston). If the Greenland ice cap and/or the Antarctic ice shelf collapses, sea levels could rise by as much as 20 ft (6 m), inundating, for example, large parts of Florida, the Gulf Coast, New Orleans and Houston. (Source: National Climate Assessment)
- **Melting ice**: Projections suggest that within the next 100 years, if not sooner, the world's glaciers will have disappeared, as will the Polar ice cap, and the huge Antarctic ice shelf. Greenland may be green in the future, and snow will have become a rare phenomenon at what are now the world's most popular ski resorts
- **Torrential downpours and more powerful storms**: While the specific conditions that produce rainfall will not change, the amount of water in the atmosphere will increase, producing

violent downpours instead of steady showers when it does rain. Hurricanes and typhoons will increase in power, and flooding will become more common. Anyone who has tried to buy storm and flood insurance in the past few years knows that the insurance industry is completely convinced that climate change is raising sea levels and increasing the number of major storms and floods.

- **Heatwaves and droughts**: Despite downpours in some places, droughts and prolonged heatwaves will become common. Rising temperatures are hardly surprising, although they do not mean that some parts of the world will not have record- cold temperatures and terrible winter storms. Global heating disturbs the entire global weather system and can shift cold upper air currents as well as hot, dry ones. Increasingly, hot, dry places will get much hotter and drier, and places that were once temperate and had regular rainfall will become hotter and drier. The string of record- high-temperature years and the record number of global droughts of the past decade will become the norm. (Source: EPA adopted from Drought Under Global Warming, A review)
- **Changing ecosystems**: As the world warms, entire ecosystems will move. Already rising temperatures at the equator have pushed such staple crops as rice north into once cooler areas, and many fish species have migrated long distances to stay in waters that are the proper temperature for them. In some cases, this may increase fishermen's catches in colder waters. In warmer waters, it may eliminate fishing. In many places, such as on the East Coast of the US, it will require fishermen to go further to reach fishing grounds. (Source: NOAA Fisheries)

  Farmers in temperate zones are finding drier conditions difficult for crops such as corn and wheat, and once prime growing zones are now threatened. Some areas may see complete ecological change. In California and on the East Coast, for example, warming will soon fundamentally change the forests; in Europe, hundreds of plant species will disappear, and hundreds more will move thousands of miles.
- **Reduced food security**: One of the most striking impacts of rising temperatures is felt in **global agriculture**, although these impacts

are felt very differently in the largely temperate developed world and in the more tropical developing world. Different crops grow best at quite specific temperatures and when those temperatures change, their productivity changes significantly. In North America, for example, rising temperatures may reduce corn and wheat productivity in the US Midwest, but expand production and productivity north of the border in Canada.

The productivity of rice, the staple food of more than one-third of the world's population, declines 10% with every 1°C increase in temperature. Past climate-induced problems have been offset by major advances in rice technology and ever larger applications of fertiliser; expectations are that in Thailand, the world's largest exporter of rice, future increases in temperatures may reduce production 25% by 2050. At the same time, global population models suggest that the developing world will add 3 billion people by 2050 and that developing world food producers must double staple food crop production by then simply to maintain current levels of food consumption.

(Source: Climate Impacts, Slideshare https://www.slideshare.net/slideshow/climate-change-causes-effects-and-solutions)

- **Pests and disease**: Rising temperatures favour agricultural pests, diseases and disease vectors. Pest populations are on the rise, and illnesses once found only in limited, tropical areas are now becoming endemic in much wider zones. In Southeast Asia, for example, where malaria had been reduced to a wet season-only disease in most areas, it is again endemic almost everywhere year-round. Likewise, dengue fever, once largely confined to tropical areas, has become endemic to the entire region. (Source: CDC) Increased temperatures also increase the reproduction rates of microbes and insects, speeding up the rate at which they develop resistance to control measures and drugs (a problem already observed with malaria in Southeast Asia).
- **Ocean acidification**: Rising temperature and rising CO_2 levels are making the world's oceans more acidic (lowering their pH). More acidic seawater damages the ability of sea creatures to make shells.

Shelled species, tiny and large, are the base of the ocean food chain, and their loss threatens the food-producing potential of the oceans.

(Source1: Environmental Protection Agency) (Source 2: National Climate Assessment)

What have we done to manage climate change? To date, the effort to manage climate change has been a matter of high-level diplomatic negotiations involving states and international organisations. All the same, in today's world, many organisations, public and private, NGOs and even startups are taking up the task of making communities aware of climate change and becoming more resilient.

CHAPTER 2

KEY SECTORS WHERE IMPACTS OF CLIMATE CHANGE ARE FELT

The impacts of climate change will affect the economy in several key sectors. Predominantly, one may look at the effects of climate change in the following four sectors.

2.1 Water sector

Water and climate change are inextricably linked. Climate change affects the world's water in complex ways. From unpredictable rainfall patterns to shrinking ice sheets, rising sea levels, severe floods and droughts - most impacts of climate change come down to water, which leads people to say that climate change is all about water.

In this sector, one may consider water-related impacts, such as floods, drought, cyclones, non-availability of clean water for communities, less water for agriculture, less water for industries, etc.

Higher average temperatures and changes in precipitation and temperature extremes affect the availability of water resources through changes in rainfall distribution, soil moisture, glacier and ice/ snowmelt, and river & groundwater flows.

These factors are expected to lead to further deterioration of water quality as well. The poor, who are the most vulnerable, are likely to be affected the most. (UN water, climate change, mainly about water)

2.2 Ecosystem sector:

An ecosystem includes all the living things (plants, animals and organisms) in a given area, interacting with each other, and with their non-living environments (weather, earth, sun, soil, climate, atmosphere).

This sector comprises industry, settlement, and society. The climate change-affected industries, settlements, and communities are often located in coastal areas and river floodplains. Wherever the economies are closely linked with climate-sensitive resources and to locations already prone to extreme weather events, the impacts are felt very harshly.

In addition, areas undergoing rapid urbanisation are vulnerable to climate impacts.

When severe weather events become more intense or more frequent, the associated ecosystems are severely affected.

2.3 Food & Agriculture sector:

The temperature changes and the frequency of droughts and floods are likely to negatively affect crop production, increasing the number of people at risk from hunger, malnutrition, displacement, and migration. In addition, the heating of ocean waters would affect fish life adversely, resulting in a lack of availability of seafood of all kinds.

2.4 Disease & Health sector:

The unpredictable climate changes are sure to alter the health condition of millions of people through the prevalence of diseases, heat stress, increased deaths, and injuries. This could happen due to heatwaves, floods, storms, fires, and droughts. In addition, increased malnutrition, diarrheal disease, and malaria in some areas will increase public health vulnerability. As a result, development goals will be threatened by long-term damage to health systems and quality of life.

The impacts on these four sectors are **highly interrelated, each affecting the others in terms of the nature of effects and intensity.** (The IPCC Fourth Assessment Report of the Working Group II "Impacts, Adaptation and Vulnerability"), (Rao and Das, 2020))

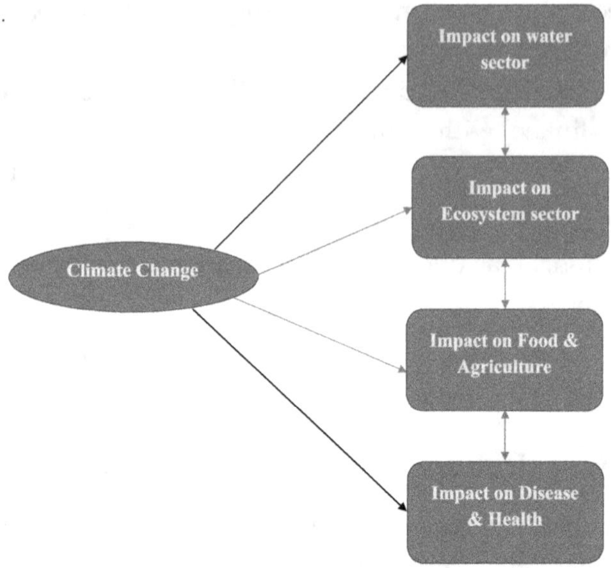

Figure 2.1 Impact of Climate Change on sectors

In addition to the four sectors, we also need to consider disasters occurring because of climate change.

2.5 Climate Change Leads to Disasters

Disasters can take the form of unprecedented floods, forest fires, extreme heat conditions, landslides and so on.

Such disasters happen when an exposed, vulnerable, and ill-prepared population or community meets with a hazardous event. Climate change will likely affect disaster risks in at least two ways. First, the risk would get accentuated with the likely increase in the frequency and severity of weather and climate hazards. Second, the risk would manifest even more through increases in the vulnerability of communities to natural hazards, mainly through ecosystem degradation, reductions in water and food availability, and changes to livelihoods. Climate change will thus add yet another dimension to the stress of environmental degradation, further reducing communities' abilities to cope with even the existing levels of weather hazards.

CHAPTER 3

STRATEGIES TO DEAL WITH CLIMATE CHANGE... MITIGATION AND ADAPTATION

3.1 In general, awareness about climate change, its causes, and its adverse effects is widely known. However, what is not so well known and not so clear is how to address these impacts strategically to help/ save communities who are suffering from floods and droughts, non-availability of food, disease affecting health, loss of housing, transport & logistics, infrastructure losses, non-availability of clean potable water, medicines and so on. To address the impacts of climate change, many worldwide forums have come up with various strategies that government and non-governmental organisations, as well as individuals, can follow.

These strategies may be broadly categorised into two types.

- ✓ a) Mitigation strategies that prevent or reduce the generation of greenhouse gases causing climate change.
- ✓ b) Adaptation strategies that help communities to adapt to the inevitable impacts of climate change which have already happened.

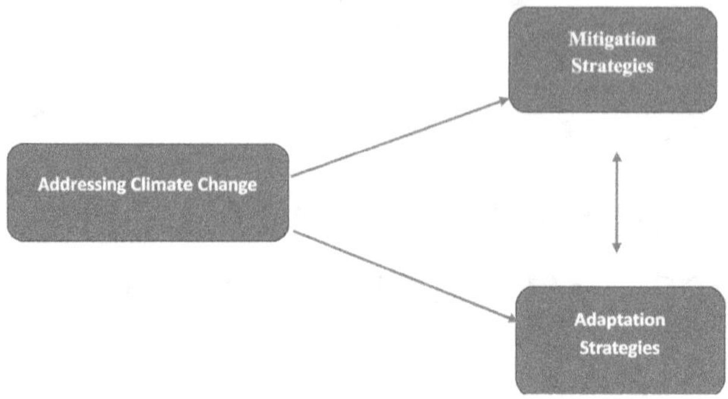

Figure 3.1 Strategies to address climate change (Source: Rao & Das, 2020)

Initial strategic frameworks focused on mitigation, aiming to reduce and possibly stabilise greenhouse gas concentrations in the atmosphere (UNFCCC 1992). However, this strategy is for the future. The increased levels of greenhouse gases which already exist **would cause and continue to cause disastrous impacts, which are unavoidable.** Thus, there is no alternative but to learn to adapt to such impacts and try to reduce the damage to the extent possible, leading to the development of adaptation strategies.

(Asian Development Bank, Climate Change ADB Programs 2007: 2009).

Recently, in research and general literature, the potential of combining both mitigation and adaptation approaches has been considered within the scientific community. It has emerged that they need to be integrated because they both influence each other. The more we mitigate, the less the need for adaptation.

(Burch and Robinson 2007, Dowlatabadi, 2007, Goklany 2007, Jones, Dettmann, Park, Rogers, and White 2007, Klein et al. 2007, and Swart and Raes, 2007).

3.2 Instances of Mitigation and Adaptation Strategies to Deal with Climate Change

In **mitigation strategy**, we are trying to reduce greenhouse gas sources and emissions and enhance greenhouse gas sinks, which help absorb greenhouse gases.

For instance, green plants have chlorophyll, which enables plants to absorb carbon dioxide and manufacture plant food in sunlight. Thus, they serve to remove carbon dioxide from the atmosphere, at least partially. Carbon dioxide is a greenhouse gas, which is not as potent as the others, but it has significantly contributed to global warming because of its abundant presence. Therefore, climate scientists welcome green plants for their contribution to reducing greenhouse gases in the atmosphere.

Other examples of mitigation actions include more efficient low-carbon furnace systems, developing new low-energy technologies for industry and transport, reducing consumption of energy-intensive products, and switching to renewable forms of energy, such as solar and wind power. Natural carbon sinks, such as forests, ecosystems, vegetation, and soils, can be encouraged to absorb carbon dioxide.

Technologies are also being developed to capture carbon dioxide at industrial sources and inject it into permanent storage, deep underground.

3.2.1 Instances of Mitigation Strategy

There has been extensive research, seminar discussions, conferences, and international forums on building awareness of mitigation strategies to reduce the generation of greenhouse gases. (Climate Change ADB Programs, 2007, Joint MDB Report 2008, Munasinghe, 2008).

These Mitigation initiatives could be at (a) the individual level and (b) the organisation level.

At the individual level, one could plant trees or prevent/protest against trees planned for cutting down. This is because trees absorb carbon dioxide, a greenhouse gas. One could also use a solar lighting system, because solar energy does not generate greenhouse gases. Other initiatives could be using energy-efficient lighting, switching off lights when unnecessary, using LED lights, using emission-free cars, reducing car trips, reducing gasoline consumption, etc.

(Reference: Climate change, ADB Programs. Strengthening Mitigation and Adaptation in Asia and the Pacific, 2007).

3.2.2 Detailing Mitigation Initiatives at the Individual Level:

Preventing/protesting against trees being cut down, because trees absorb CO_2, a greenhouse gas.

Planting trees helps because trees absorb CO_2.

Using a solar lighting system, solar energy is a form of clean energy, so no emissions.

Using energy-efficient lighting results in less GHG generation.

Switching off the lights when additional light is not needed in the room.

Using LED-based decorative lighting generates fewer GHGs. (LED is a light-emitting diode).

Using emission-free cars.

Reducing car trips and reducing the consumption of gasoline.

Reducing food waste because waste food generates significant GHG emissions, accounting for **8-10% of total global GHG emissions.**

3.2.3 Mitigation Possibilities at Organisation Level:

Using a climate change-friendly production system and cooling system, which minimises the generation of greenhouse gases.

Using renewable energy, solar and other alternative energy in the production systems.

Using boilers which do not produce greenhouse gases.

Using input materials which do not produce greenhouse gases.

Encouraging suppliers to use cleaner production systems.

Using transportation and distribution systems which minimise the production of greenhouse gases.

Using systems to capture methane gas from landfills, so that methane does not go to the atmosphere.

Using cleaning agents which do not generate greenhouse gases.

Reducing industrial waste, which is a significant source of GHG emissions, primarily producing methane, carbon dioxide and nitrous oxide. These emissions arise from anaerobic decomposition in landfills, industrial wastewater treatment, and waste incineration, contributing to about 3.4% to 4% of global or national emissions.

Ensuring that buildings have adequate insulation in order to minimise the energy needed for cooling in summer.

Ensuring that natural light is used where possible in order to reduce demand for lighting.

Using solar energy in production.

3.2.4 To give an example of mitigation initiatives in the FMCG Industry in India

(FMCG: Fast-Moving Consumer Goods)

In this industry, there have been significant efforts to produce products which do not generate greenhouse gases upon use by customers. (Patil, 2016)

On the warehousing front, the warehouse contractors are advised to work on better insulation of the walls of the warehouse so that it leads to less energy consumption.

There have been many initiatives to replace the lighting systems with energy-efficient lights. LEDs, smart and movement-sensing lighting systems are expected to lead to energy savings of up to 90%.

Since transportation is a major cost component in the overall logistics cost of an FMCG company, organisations are working on improving the fuel efficiency of the trucks. They are trying to reduce the idling time of trucks on transit. They are also working on pickup strategies for trucks while delivering.

A paper and packaging conglomerate has taken a step further, wherein they have been engaging local communities to grow trees according to the demands of the local ecosystem, which they later consume in their paper and packaging plants.

A major food company has begun using spent coffee grounds and biogas from wastewater to fuel its boilers and enlisted its entire supply chain to reduce energy consumption and cost within a global logistics initiative.

Also, the companies are trying to apply mitigation across the whole supply chain and not only to some of its parts. Thus, the value delivery of the supply chain will be greatly enhanced in terms of tangible and intangible mitigation benefits, addressing major issues in climate change.

(Verma, Malhotra and Rao, 2017)

3.3 Adaptation Strategy

The second task in responding to climate change is to manage its impacts for the damage which has already been done. This is **adaptation strategy**. Future impacts on the environment and society are now inevitable due to the amount of greenhouse gases already in the atmosphere from decades of industrial and other human activity, and the added amounts that will happen from continuing emissions over the next few decades.

In adaptation strategy, we are trying to prepare and make adjustments in natural or human systems in response to actual or expected climatic reactions or their adverse effects. Examples of adaptation include preparing risk assessments, building dikes to act as protection from floods, protecting & restoring ecosystems, improving agricultural methods, managing water resources, building settlements in safe zones, developing early warning systems, instituting better building designs, improving insurance coverage, and developing social safety nets. These measures are intrinsically linked to sustainable development, as they reduce the risk to lives and livelihoods and increase the resilience of communities to all hazards.

Ideally, adaptation and mitigation should be considered jointly, as some adaptation measures can reduce greenhouse gas emissions as well. Conversely, some mitigation measures can be planned to help reduce and not inadvertently exacerbate disaster risks. (Climate Change and Disaster Risk Reduction 2008).

3.3.1 Instances of Adaptation Strategy

Adaptation in the context of climate change refers to how communities, groups of people, sectors, regions or even countries develop systems to better cope with the impacts of climate change. (Brooks, 2005, Smit et al. 2003, Pielke, 2007).

Adaptation strategies are often specific to local conditions and are locally relevant action plans aimed at reducing climate-related risks and enhancing resilience.

These action plans are community-based development projects promoting information sharing, developing early warning systems (to warn of an impending calamity) and preparedness plans, developing more diverse crop strains that can withstand a variety of conditions (heat, drought, salt, etc.), increasing storage capacity for fresh water by building reservoirs or by recharging aquifers, improving public health infrastructure, and bolstering disease surveillance.

3.3.2 Various Aspects of Adaptation Strategy

Adaptation may also be defined as the adjustments in individual groups and institutional behaviour to reduce society's vulnerability to climate change.

The adaptation process can be anticipatory or reactive and could be autonomous or planned.

(Fankhauser et al., 1999; Smit et al., 2000). Governmental agencies, municipalities, and NGOs have implemented many adaptation strategies in general.

As mentioned earlier, adaptation strategies are specific to local conditions; they consist of locally appropriate action plans to reduce climate-related risks. These action plans are community-based development projects to promote information sharing, develop early warning systems (warn of an impending calamity) and preparedness plans.

3.3.3 Specific Initiatives in Adaptation

*Developing more diverse crop strains that can withstand various conditions (heat, drought, salt, etc.),

*Bolstering social capital and resilience,

*Increasing storage capacity for freshwater by building reservoirs or recharging aquifers, improving public health infrastructure, and bolstering disease surveillance. These strategies would be valuable regardless of the exact impacts of climate change at a particular time or location. (GEF 2009).

Adaptation strategies should be developed, looking at the precise nature of vulnerability of the target communities, who are in danger of climate change impacts. The strategies should be suited to help them to best adapt to address the impacts. (Kelly and Adger, 2000; Downing, 2001; Turner et al., 2003; Smit and Pilifosova, 2003; Yohe et al., 2003; Adger, 2006).

3.3.4 Adaptation in the Indian Scenario

In India, climate change impacts have been one of the main causes of water hazards over the last two decades.

Here, the major rivers get their water flow from the snow and glaciers, and melt in the seasons before the summer monsoon. If climate change continues to raise the temperature in the atmosphere, the river flow in the Himalayan basins may initially increase, but when the reservoirs of snow and ice continue to deplete, the variability of downstream run-off will increase and water flow will subsequently reduce. This will seriously affect the livelihood of people in the river basin, including fisheries and agriculture (Grail Research, 2009; Wick & Wittgren, 2009).

To address many such water-related threats in India, coping mechanisms in the form of adaptation strategies are urgently needed.

Some of the areas in India where adaptation will be most appropriate, either by the public sector or by the private sector, are outlined below.

Areas where water-related Adaptation will be needed in India:

- Due to warming temperatures as a result of climate change, there will be higher water demands from the population,
- Due to a change in the hydrological cycle, there will be too much rain in some months (resulting in floods) and too little rain in others (resulting in droughts),
- In India, more areas and hence more population will fall under severe water stress,
- Crop production will suffer due to water shortage and increasing temperature,
- Shortage in water supply will affect power generation and other industries that need large volumes of water,
- There would be increased disease and heat stress due to water shortage,
- Mass migration in search of water will become commonplace,
- Irrigation, consuming 70% of total water needs, will suffer drastically,
- Water-borne and water-washed diseases will increase,
- Water-related diseases will increase,
- Natural habitats of water-borne diseases are likely to expand,
- Brackish water would affect aquaculture.

3.4 Possible Adaptation Strategies for Water sector in the above context

In many of the above areas, there exist options to go for climate change adaptation relating to water. In most such initiatives, the private sector can be involved in a significant way.

Summarising Adaptation to Reduce Water Demand in the Face of Shortage

In the face of water shortage, initiatives can help to:

- Enhance industry investment in recycling and treatment of industrial wastewater,
- Reduce water use,
- Incorporate efficient water usage practices,
- Promote conservation of water,
- Practice rainwater harvesting
- Invest in wastewater treatment plants to reduce water demand,
- Start incorporating 3R: reduce, reuse, recycle,
- Incorporate sewage water management and recycling,
- Start monitoring of water consumption,
- Promote Optimisation of operations such as cooling tower operations.

Source (Rao, 2020)

CHAPTER 4

HOW CAN CORPORATE MANAGERS/ PRIVATE SECTOR HELP ADDRESS CLIMATE CHANGE CONCERNS?

4.1 Given the magnitude of help required to protect populations from climate change, corporate managers could come forward and use their managerial skills and expertise to take up initiatives to address climate change. With its unique expertise and ability to innovate, the private sector could augment government efforts to build capacity for the communities affected by such calamities to survive the adversities caused by climate change. Corporate managers throughout the industry have been effective in planning and carrying out successful projects to bring about efficiency from the organisational perspective. If they take it upon themselves to address an impending climactic disaster, they can be equally effective in enhancing resilience and minimising vulnerability.

(http://unfccc.int/adaptation/workstreams/nairobi_work_programme/items/4623.php).

There is also another important reason why the private sector should address climate change-related initiatives. The private sector generates a tremendous amount of greenhouse gases through its operations. Not only do corporations have a significant climate footprint, but the impact of climate change on the business landscape is already noticeable. (Patenaude, 2011). Operating manufacturing setups use coal and other fossil fuels, use unlimited resources like water and electricity, deploy massive industrial air conditioning units that generate greenhouse gases, use non-renewable lighting and employ non-environment-friendly transportation & distribution systems. Thus, the private sector has been a significant contributor to global warming and climate change. So, it is crucial that private sector organisations be made mindful of the perils of their activities and devise systems/ activities and use materials that minimise greenhouse gas generation.

So far, the involvement of the private sector in the climate change issue has been mostly in the realm of mitigation strategies and carbon trading. A company reduces the emission level, equates the reduction to what is called a certified emissions reduction (CER), and sells it to others that need to meet CER targets. However, though emission saving is achieved at the world level, carbon trading does not necessarily lead to adaptation or reduction in vulnerability for the poor, which is an urgent need in today's world. (McCarthy et al., 2001). Also, after policies regarding mitigation and adaptation have been designed, it would be necessary to translate the policies into action plans, implement them in real terms and bring about an actual reduction in climate change. (Halady and Rao, 2010).

Preclude to this would be enhancing the awareness of the impacts of climate change and initiatives to address the challenge (Burgess et al., 1998). For the private sector, especially for small and medium enterprises (SMEs), it is essential to know what to do in the face of climate change calamities and how to make their businesses resilient. While climate change poses several risks to vulnerable communities and enterprises worldwide, many opportunities are unfolding for private companies to implement actions towards reducing risks to their business operations and investing in adaptation action in vulnerable regions sustainably and profitably.

Adaptation activities for corporate organisations may thus relate both to ensuring the resilience of business operations and the provision of technologies or services that assist in the adaptation for vulnerable communities.

(http://unfccc.int/adaptation/workstreams/nairobi_work_programme/items/6547.php).

In order to act, corporate managers need a thorough knowledge base and awareness to take up planning and preparedness effectively and efficiently. Such awareness and know-how could be imparted to the managers through continuous managerial publications, seminars, and discussion forums. (Halady & Rao, 2010)

Over the last few years, there have also been numerous research initiatives to explain the gap between environmental awareness, environmental knowledge, and the actual display of pro-environmental behaviour (Kollmus and Agyeman, 2002). Using the learning from such research, one may observe that knowledge and awareness, as well as other variables such as availability of funding and resources, societal pressures, and an individual or organisation's capability to take up the challenge, can lead to initiatives to address climate change.

4.2 Empirical research showing awareness is the key to corporate initiatives

In an empirical research study conducted with a section of Indian managers on their awareness of climate change impacts and their pro-climate change behaviour, it emerged that there was a significant association between awareness and initiative to behaviour change. (Halady and Rao, 2010). Thus, once awareness sets in, initiatives towards addressing the adverse effects of climate change will also start to happen. In particular, awareness of the health impacts of climate change was found to significantly lead to initiatives to act on climate-friendly behaviour.

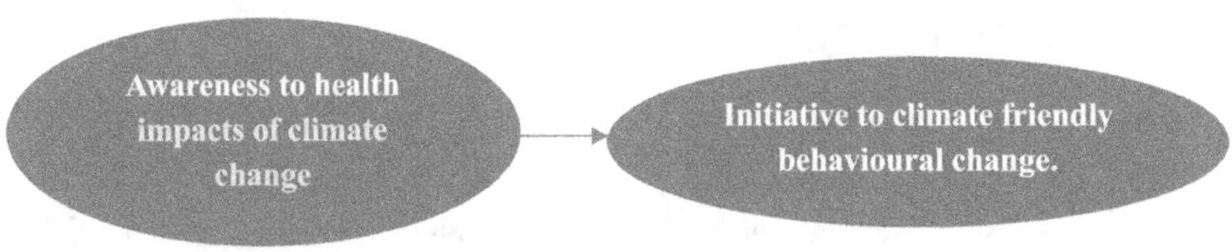

Figure 4.1 Awareness leads to behavioural change: Findings from empirical research (Halady and Rao, 2010)

A subsequent empirical study was conducted in Southern India to evaluate/measure the linkages between awareness of climate change, its impact on the poor, and the willingness of the private sector to act on adaptation and mitigation strategies. Again, it emerged that there was a significant linkage between awareness and the willingness of the corporate executives to support adaptation strategies. (Rao & Thamizvanan, 2014).

In 2020, another empirical research supported the conceptual hypothesis that awareness, especially in health impacts, led to a significantly higher propensity to initiatives to address climate change impacts (Rao, 2020). Awareness encompasses not just the existence of impacts of climate change but also that possible initiatives exist. Executives need to be made aware of how they can make a difference by learning about the roadmaps, existing initiatives, and how they can implement them. (Rao, 2020).

Corporate managers throughout the industry have been effective in planning and carrying out successful projects to bring about efficiency from the organisational perspective. If they take it upon themselves to address an impending climactic disaster, they can be equally effective in enhancing resilience and minimising vulnerability.

(http://unfccc.int/adaptation/workstreams/nairobi_work_programme/items/4623.php).

Adaptation activities for corporate organisations may thus relate either to ensuring the resilience of business operations or the provision of technologies or services that assist in the adaptation for vulnerable communities.

(http://unfccc.int/adaptation/workstreams/nairobi_work_programme/items/6547.php).

In addition to corporate managers' need for a thorough knowledge base and awareness to take up planning and preparedness effectively and efficiently, they also need to know possible strategies to address the impacts. Thus, they need to know the Mitigation strategies and Adaptation strategies thoroughly. Such know-how could be imparted to the managers through continuous managerial publications, seminars, and discussion forums. (Halady & Rao, 2010)

Over the last few years, there have also been numerous research initiatives to explain the gap between environmental awareness, environmental knowledge, and the actual display of pro-environmental behaviour (Kollmus and Agyeman, 2002). Using the learning from such research, one may observe that knowledge and awareness, availability of funding and resources, societal pressures, and an individual or organisation's capability to take up the challenge, can lead to initiatives to address climate change. (Rao, 2020)

4.3 Research on Existing Industry Initiatives to Address Water Impact of Climate Change

In this research, climate change impacts on water specifically, were considered.

Climate change impacts the availability of clean water for communities, water for agriculture, and industrial use, among other impacts. Industry investment in recycling and treatment of industrial wastewater to reduce water demand has been in existence and they surely help.

Also, in practice, they are incorporating efficient water usage practices, water conservation, rainwater harvesting, monitoring water consumption, and wastewater treatment plants to reduce water demand. These are of course helpful. In addition, optimisation of operations such as cooling tower operations would help address the water shortage impact of climate change.

Strategy Framework for Corporate Managers to Address Climate Change

The journey from a willingness to help to the actual implementation of initiatives sometimes requires a structured framework. This framework can help corporate managers and business leaders identify and develop strategies to address the impacts of climate change in mitigation and adaptation, integrate these into planning processes, and influence future developmental activities. Consequently, the organisations and the communities they support become more resilient to climate change and variability.

The framework builds and elaborates on the work done by many countries since 2002 on Integrated Water Resources Management (IWRM) plans.

(EU, Water Initiative Finance Working Group. 2010. Strategic Financial Planning for Water Supply and Sanitation in Africa. May 2010. Available at: www.euwi.net/wg/finance)

There are two phases in the framework.

In the **first phase**, the executives need to be aware of the entire climate change phenomenon, the reasons, impacts on our world, and implications for industry and communities. This phase, **building awareness phase**, can be achieved with the help of climate change experts and specialists.

In the **second phase, the implementation phase**, executives need to be ready to implement feasible strategies to address climate change impacts through mitigation and adaptation.

To identify the climate change impacts on all stakeholders, a good starting point could be the organisation's integrated supply chain.

Further to the empirical research by Rao & Thamizvanan (2014), and by Halady & Rao (2010), surveys were conducted on sampling frames of practising Indian managers, considering the research question as to what kind of awareness leads managers in the private sector to take up the climate change cause. The research concluded that Awareness to Disease and Morbidity, Lack of food and water and Impact on nature would significantly lead to Adaptation strategy as taken up by managers, whereas Awareness to Impact on nature would significantly lead also to Mitigation Strategy.

The diagram below gives the final structural model as arrived at by the research. Here, the bold line indicates a significant link and the dotted line indicates a link not significant yet.

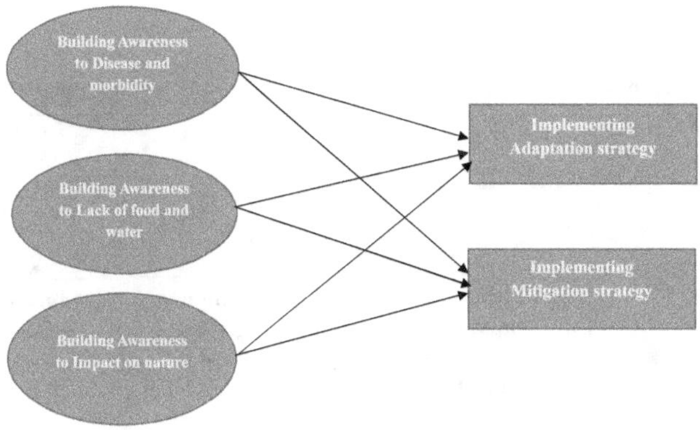

Figure 4.2 Building awareness leads to Strategy to implement Adaptation and Mitigation.

Mitigation strategy is really a far-reaching approach involving taking steps to prevent/ reduce the onslaught of climate change in the future. In contrast, adaptation addresses the impacts happening now, like flash floods, hurricanes, tsunamis, droughts, prevalence of disease and so on. Thus, adaptation is going to take effect in current times, which is urgently needed in today's world. This is where the private sector is needed to come forward and the empirical research validates that awareness to the climate change impacts would indeed lead to this action plan significantly.

4.4 Some Corporate Initiatives to address Climate Change in different industries.

4.4.1 Semiconductor Industry

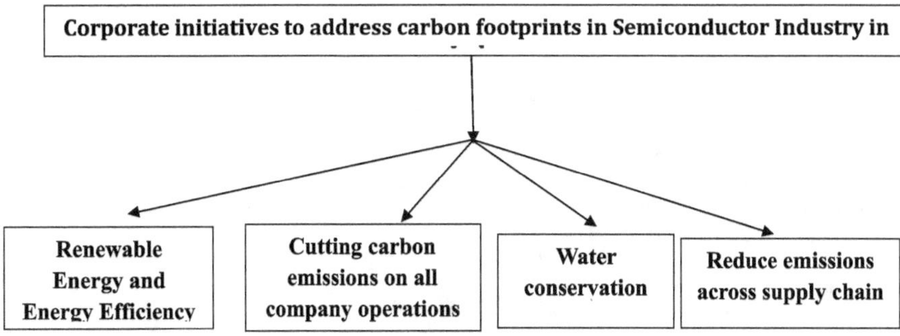

Figure 4.3 Corporate Initiatives in the Semiconductor Industry

4.4.2 Textile Industry

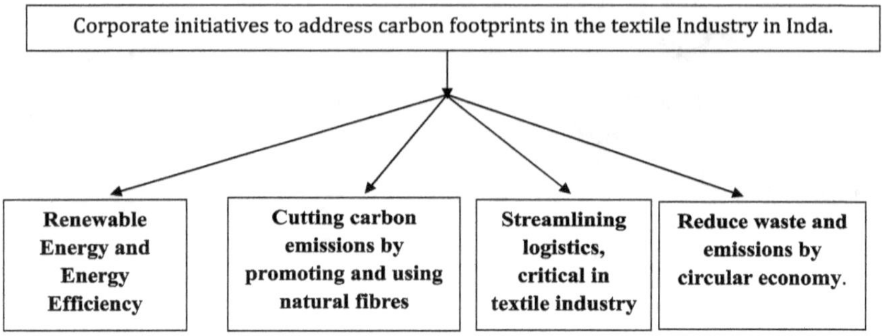

Figure 4.4 Corporate Initiatives in the textile industry

4.4.3 FMCG Industry: (Fast-Moving Consumer Goods)

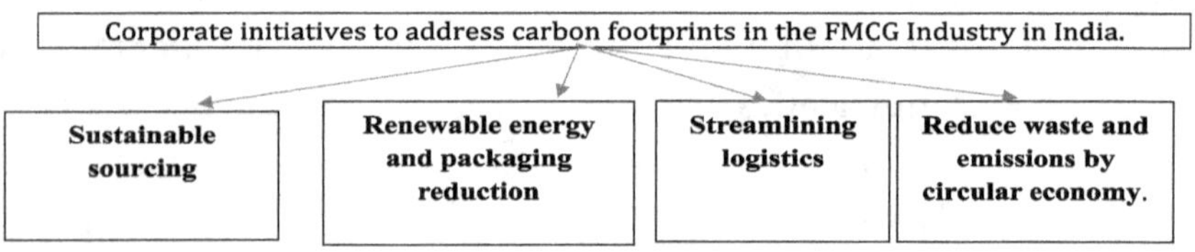

Figure 4.5 Corporate Initiatives in the FMCG industry

In the following chapters, we will discuss the climate change impacts for each of these sectors.

CHAPTER 5

CLIMATE CHANGE IMPACTS ON THE WATER SECTOR

5.1 Climate change has always had a huge impact on water issues. Climate change and water have been inextricably linked. It is known that climate change affects our world's water in many complex ways. On one hand, there are unpredictable rainfall patterns which lead to rising water levels in rivers, lakes and the sea, often leading to floods. On the other hand, there is often a huge scarcity of clean water too. Shrinking ice sheets and droughts affect agriculture, industry and mostly human livelihood in villages and cities. Thus, many of the impacts of climate change come down to water, **which leads people to say that climate change is all about water.**

In this section on the water sector, we consider many water-related impacts, such as floods (too much water), drought (less water), cyclones, non-availability of clean water for communities, for agriculture, for industries, etc. Sometimes there is too much water. Sometimes there is too little clean water for all.

Higher average temperatures due to climate change and changes in precipitation and temperature extremes affect the availability of water resources through changes in rainfall distribution, soil moisture, glacier and ice/snowmelt, and river & groundwater flows. These factors are expected to lead to further deterioration of water quality as well. The poor, who are the most vulnerable, are likely to be affected the most. (UN water)

Climate change is exacerbating both water scarcity and water-related hazards (such as floods and droughts), as rising temperatures disrupt precipitation patterns and the entire water cycle (UNICEF).

5.2 Climate Change Impact on Water in Agriculture

Water is the prime channel through which the impacts of climate change on the world's ecosystems and livelihoods will be felt. Climate change has the potential to affect every element in the water cycle (UN-Water, 2010). Agriculture will be affected by increased evaporative demand, changes in the amount of rainfall and rainfall patterns, and variations in river run-off and groundwater recharge, which are the two major sources of water for irrigation

5.3 Renewable Surface Water and Groundwater Resources

Precipitation and evaporation are the main climatic phenomena affecting freshwater resources. Climate models predict decreases in renewable water resources in some regions and increases in others. The decrease of renewable surface water and groundwater resources in dry subtropical regions will intensify competition for water by different users (e.g. agriculture, ecosystems, settlements, industry, and energy production). This will have an impact on regional water, energy and food security. How changes in vegetation brought about by increasing concentrations of greenhouse gas in the atmosphere will affect water resources and irrigation requirements remains uncertain.

5.4 Water Quality

There is high agreement that climate change negatively impacts freshwater ecosystems by changing streamflow and water quality. Climate change affects the quality of water through a complex set of natural processes and anthropogenic factors.

There is medium evidence and high agreement that climate change will likely reduce raw water quality, which will pose risks to drinking water quality with conventional treatment. The sources of these risks are increased temperatures, higher levels of sediment, nutrient and pollutant loading due to heavy rainfall, reduced dilution of pollutants during droughts, and disruption of treatment facilities during floods.

5.5 Streamflow Seasonality

In regions with snowfall, climate change has already altered the seasonality of streamflow. There is robust evidence and high agreement that this variability will increase. Except in very cold regions, warming in the last decades has reduced springs and, maximum snow depth and brought forward the maximum of snowmelt discharge. Smaller snowmelt floods, increased winter flows, and reduced summer flows have all been noted and are to be addressed. River ice in Arctic rivers has been observed to break up earlier.

5.6 Groundwater Recharge

The impact of climate change on groundwater recharge is difficult to predict. Groundwater recharge is the hydrologic process where surface water moves downward through soil and rock to replenish underground aquifers. Changes in precipitation intensity will affect the amount of total run-off that recharges groundwater. In humid areas, increased precipitation intensity may decrease groundwater recharge because the infiltration capacity of the soil will be insufficient. In semi-arid areas, increased precipitation intensity may increase groundwater recharge due to the faster rate of percolation through the root zone, which will reduce evapotranspiration. Evapotranspiration is a beneficial process and should not be reduced.

Evapotranspiration (ET) is the combined process of water transferring from the land surface to the atmosphere via evaporation (from soil, water bodies) and transpiration (released by plants).

5.7 Crop Water Demand

Changes in climate will affect the water demand of crops grown in both irrigated and rain-fed systems. An increase in temperatures will trigger a higher demand for water for evapotranspiration by crops and natural vegetation, which will lead to more rapid depletion of soil moisture. This scenario, combined with changes in rainfall patterns, may lead to more frequent crop failures. In general, projections show that the water demand to produce a given amount of food on either irrigated or rain-fed systems will increase in many regions due to climate change. The differences in yield and yield variability between rain-fed and irrigated land may also increase with changes in climate and greater climate variability.

5.8 Sea Level

It is virtually certain that sea level will continue to rise during the 21st century and beyond. For all scenarios, the rate of 21st-century global mean sea level rise is very likely to exceed the average rate during the 20th century. The expected rise in sea levels will affect agriculture in coastal areas, particularly river deltas. Higher sea levels combined with variations in the distribution of run-off and more frequent floods in upstream areas will result in an increased incidence of flooding and saltwater intrusion in estuaries and aquifers. This will affect some of the world's most productive agricultural areas.

5.9 Climate Change Impacts on Floods and Droughts

We know that climate change is projected to alter the frequency and magnitude of floods and droughts. The extent of impact is expected to vary from region to region. Though there is limited evidence, there is high agreement that floods will increase over more than half of the globe, particularly in central and eastern Siberia, parts of Asia, including India, tropical Africa, and northern South America. In some regions, including southern Europe and West Africa, meteorological droughts, which are characterised by significantly reduced rainfall, and agricultural droughts, which are characterised by significant declines in soil moisture that affect crop production, have become more frequent since 1950 (Seneviratne et al., 2012). Climate change may increase the frequency and length of droughts by the end of the 21st century. However, it is not clear what these rainfall and soil moisture deficits may mean for prolonged reductions of streamflow and lake and groundwater levels.

5.10 Water-related Hazards

Because of climate change, extreme weather events such as floods and droughts have a greater frequency now and have become more severe (IPCC). In addition, rising global temperature continuously increases the moisture the atmosphere can hold, leading to more storms and heavy rains. Also, droughts are more intense now, involving dry spells as more water evaporates from the land on account of higher temperatures. (World Bank)

Research has projected a further increase in drought, flood risks and related social damages with warming of the global temperature. (IPCC).

In addition, heavy precipitation events will occur with more frequency in the next century, with more rain-generated floods.

Referring to disasters on account of climate change, water-related disasters have been predicted with great risk to humanity over the next decades. (World Bank).

Since 2012, the water crisis has been consistently considered as one of the most important threats facing the world, having the highest potential damage (Berggren, 2019). The crisis is due to extreme weather conditions caused by climate change and other natural disasters. In many parts of the world, water-related climate change impacts are being encountered in the form of more severe and more frequent droughts and floods. The factors like higher average temperatures & changes in precipitation are affecting the availability of water resources through changes in rainfall distribution, soil moisture, glacier and ice/ snow melt, and river and groundwater flows.

5.11 Strategies to Address Water-related Impacts

For **the mitigation aspect** of water-related strategy, the water sector has to focus on water harvesting and the productive use of rainfall to overcome water shortages at small scales. The water management strategy should focus on:

*Preventing pollution of all kinds,

*Treating inevitable waste streams,

* Reusing wastewater flows,

*Campaigning for a change towards consumption patterns with a lower water footprint,

*Developing water footprint benchmarks for crops and products that reflect reasonable levels of water consumption per unit of production and work towards achieving those benchmarks by focusing

on smart and efficient irrigation scheduling, improved soil and crop management,

*Campaigning to not tap into groundwater resources, which are to be used only in urgent times, in low amounts and infrequently, etc.

(http://www.globalwaterforum.org/2016/01/04/strategies-for-countries -to-mitigate-the-risks-of-extreme-water-scarcity-and-dependency-in-a-sustainable-way/)

For **adaptation**, one may note that throughout the ages, communities and societies have always had to cope with various kinds of natural hazards such as floods, droughts, forest fires, earthquakes, and spread of diseases. This coping with nature is essentially adaptation of whole societies to natural phenomena. Such adaptation strategies have taken the form of seasonal migration of communities in search of water in the face of drought, agricultural civilisations inventing techniques for storing water, transfer and irrigation, constructing embankments and dikes and even lowering lakes to protect cities, agricultural land and populations from flooding (Wilk & Wittgren, 2009).

Thus, there do exist many different approaches to address the water crisis situation today. To streamline and optimise this enormous work, the industry and the private sector, in general, should be encouraged to enhance efficiency, competence and strategic thinking for delivery of action plans that help our country develop resilience in all aspects of water challenges in situations of floods, water shortage, disease and unsanitary conditions.

All the same, the perspective and awareness of the private sector under the whole umbrella of possible initiatives, has not been widely assessed yet. Thus, to address the climate change challenges in the water sector, awareness of the private sector needs to be improved with regard to the possible initiatives for addressing the impacts of the water situation, both inward-looking and outward-looking.

5.12 How Industry Can Help in Water Impacts

The industry, communities, and water sector must adapt to these changing climatic conditions by developing improved water management approaches that will reduce pressure on already stressed systems. Thus, there will be an urgent need to develop and implement technologies and policies that will reduce and combat greenhouse gas emissions on one side and develop adaptation mechanisms to help communities survive against sudden and intense floods, droughts, diseases and lack of food. In this context, one may consider the **two interrelated and often co-existing strategies to address water management,** which may be adopted by government agencies, NGOs, private sector organisations and communities to help build resilience against the impending water disasters. These two strategies may be broadly classified as "Inward" and "Outward."

Inward and Outward Corporate Initiatives to Address the Water Crisis Situation

Corporate initiatives to address the water crisis may be broadly categorised as (i) inward, when companies look inwards into their own operations, and (ii) outward, when companies look beyond the walls of the factory to help communities and the surrounding environment become water efficient. Many industries use huge volumes of water as a solvent and cleaning agent, to cool industrial and energy generation processes, to dilute contaminants, to extract fossil fuels, and as a key ingredient in many products, among many other uses. However, since water scarcity has become a concern in India, many businesses are making efforts to promote and invest in improved water resource management (CDP, 2012; Ceres, 2009). This is based on their growing understanding that water risks are sometimes caused by their own water use and pollution, hence necessitating efficient water management inside the company. For example, when a company releases harmful effluents into the water bodies surrounding the company, the communities outside the company are also drastically affected.

Some inward corporate water initiatives comprise:

1. investment in recycling and treatment of industrial wastewater,
2. incorporation of efficient water usage practices in company operations,

3. investment in rainwater harvesting facilities,
4. recycling of industrial wastewater discharge internal to company,
5. investment in wastewater treatment plants so the polluted water does not go out,
6. incorporation of 3R: reduce, reuse, recycle,
7. incorporation of sewage water management and recycling of the same,
8. monitoring of water consumption in operations,
9. optimisation of cooling tower operations,
10. setting up water conservation equipment,
11. putting up used water cleaning and recirculation systems,
12. optimisation of inbound logistics so as to conserve water,
13. holding awareness seminars on water conservation for employees,
14. optimisation of processes to reduce effluent wastewater,
15. prevention of pollution at the source,
16. striving to make the industry zero-water industry,
17. promoting green building concept on water conservation programme.

    (Grail research, 2009).

Some outward corporate water initiatives comprise:

1. ensuring that effluent wastewater and other pollutants do not enter any of the water bodies,
2. help in providing irrigation water and increasing agricultural production in the community,
3. watershed development as a community initiative,
4. providing drinking water and purification systems in a community,
5. helping communities outside the factory to access clean water.

(addressing the cause and adaptation) (addressing the impacts which can no longer be avoided) (Mitchell & van Aalst, 2008; Rao et al., 2016; Halady & Rao, 2010).

5.12 Indian Context on Water

In the years to come, India will be highly vulnerable to the impacts of rising sea levels, floods and droughts, water pollution, and the associated health hazards. In addition to the severe effects of water-related concerns on health and heat stress, water scarcity problems would also impact generation of electricity, both in hydel and thermal power sectors, industrial production, which often requires large volumes of water, and irrigation, which again is highly water-based. To address such severe concerns in the realm of the unavailability of water in desired quantity and quality, many industry initiatives have been implemented. However, the challenge is huge and more industry participation in this endeavour is needed in India. Private sector participation would be a relevant and much-needed initiative to the cause of water management in the country.

Like many other developing countries, water stress in India is already high, and climate change adds even more urgency for action.

In developing countries, like India, it is often a challenge to provide clean water of the desired quantity and quality at the desired place. This is especially true in monsoon climates where 70-90% of the annual rain falls in just 3-4 months. So, either there is too much water and often floods in the wet season, or too little water and often droughts in the dry season. Also, at times, enough water may be available but the quality may be unacceptable (Jain, 2012).

Parikh (2013) says that India's water crisis, aggravated by climate change, is rooted in three causes. The **first cause is insufficient water per person** as a result of population growth. With a population of 1.2 billion, according to the 2011 census, we have only 1,000 cubic metres of water per person, far less than the required standard of 1700 cubic metres per person per year. Actually, a country is considered water-stressed if it has less than 1,700 cubic metres per person per year. The insufficient water is felt even more because of climate change.

The **second cause** for the water crisis is **poor water quality** as a result of lack of proper investment in water-treatment facilities in urban India. Water in most rivers in India is largely not fit for drinking, and in many stretches not even fit for bathing. Despite some initiatives being taken to clean up rivers, much of the river water remains polluted with a high coliform count at many places (Parikh, 2013). [Coliforms are a group of common bacteria found in soil, water, plants, and the intestines of warm-blooded animals and humans].

The **third cause** for the water crisis is dwindling groundwater supplies due to over-extraction by farmers and reduced fresh water supply due to climate change. Also, **industrial pollution is seeping into the groundwater**. All the same, one of the most significant causes of water pollution is industry dumping their effluent waste in the rivers, lakes, canals and other water bodies in an uncontrollable manner. The effluents contaminate the water so much that the people residing around the polluted water bodies develop skin diseases, stomach problems and other health issues. (http:// www.deccanchronicle.com/nation/ current-affairs/300717/ more-bellandur-lakes-in-making.html).

In many parts of the world today, particularly in emerging economies where the spread of industrial and agricultural activity has become very high, the water scarcity situation is becoming a problem because the consumption of water is much larger than what the local aquifers and river basins can hold.

Climate Change → reduced supply of fresh water → Water scarcity

Climate Change → floods and excess rainfall → danger to communities.

5.13 Corporate Initiatives in Water Management in India

In many situations, water consumption exceeds water availability (which is reducing perpetually on account of climate change) by a ratio of more than three to one (www. ceres.org/resources/reports/ aqua-gauge, page 41). Today's industry uses fresh water directly for its manufacturing activities, and relies on water indirectly too in the processing of supplies from the suppliers in the inbound supply chain. It uses water as a solvent and cleaning agent, for cooling purposes in industrial and energy generation processes, to wash away contaminants, for irrigation, and to extract fossil fuels. It is used as a key ingredient in many products too. Because of large-scale water scarcity, lack of clean water, and inefficient water management systems, many businesses are strategically inclined to promote and invest in improved water resource management (CDP 2012, Ceres, 2009). These strategies are based on their growing understanding that water risks are caused not only by a company's own water use and pollution, but also by the watershed context in the environment in which the company operates.

Thus, water risks can be mitigated if the industry addresses water concerns both inside and outside of its organisation. As part of these concerns, industry also needs to be aware that if they produce effluent water waste, they would need to mitigate the effluents before releasing wastewater outside, in the local water bodies. This can be achieved by putting up water-treatment plants on their premises. In this context, the CEO Water Mandate (2014) proposes the business case for sustainable water management and develops a system of good practice guidance to identify and encourage integrated corporate water stewardship. The mandate argues that companies can not only act significantly to achieve more sustainable and effective water management in their own operations, but that in many areas their participation may prove to be vital, particularly in emerging markets where government capacity and resources are limited.

The participation of companies can contribute to uplifting the lifestyles of communities too, in a significant way. These days, corporate social responsibility (CSR) has been widely accepted. In the context of water management, CSR could provide major solutions to the water crisis situation in many ways.

On one side, a CSR programme is relevant for companies that use lots of water in their operations. Here, it is related to water efficiency that goes along with a responsibility to ensure their surrounding communities will not suffer from water shortages because of the company's operations. On the other side, water provision can be seen as a business opportunity that provides a solution for the community; hence it refers to social entrepreneurship (Triwadiantini, 2010).

5.14 Mitigation in the Water Sector in India

To mitigate water issues, key strategies include: water conservation through demand management, efficient irrigation practices, rainwater harvesting, wastewater treatment and reuse, watershed management, pollution reduction at source, and restoration of natural water retention areas like wetlands; all aimed at reducing water usage, minimising pollution, and enhancing water availability in the face of increasing demand and climate change.

Key points to consider:

- **Demand management:**

  Implementing policies to encourage responsible water use, like fixing leaks, using water-efficient appliances, and public awareness campaigns.

- **Efficient irrigation:**

  Utilising technologies like drip irrigation systems to minimise water loss in agriculture.

- **Rainwater harvesting:**

  Collecting rainwater from rooftops and storing it for later use, especially in areas with seasonal rainfall.

- **Wastewater treatment and reuse:**

  Treating wastewater to a standard suitable for reuse in industrial processes or irrigation, thereby reducing fresh water demand.

- **Watershed management:**

  Protecting and restoring natural landscapes within a watershed to improve water quality and regulate water flow.

- **Pollution reduction at source:**

  Addressing industrial and agricultural run-off by implementing stricter regulations and promoting best practices to minimise pollutants entering water bodies.

- **Wetland restoration:**

  Re-establishing natural wetlands to act as natural filters and regulate water levels.

5.14.1 Examples of Mitigation Initiatives in the Water Sector

India is the world's third-largest methane emitter, with agriculture as its primary source, contributing 12% of global agricultural methane. Major contributors are livestock enteric fermentation (48%) and flooded rice cultivation (15%), with emissions projected to rise by 7.8% by 2030. (FAO, 2014)

1. **How water management can help climate change mitigation**

   Over the past 50 years, greenhouse gas emissions resulting from agriculture, forestry and other land uses have nearly doubled, and projections suggest a further increase by 2050 (FAO, 2014).

   So, climate change mitigation practices are urgent. Land management practices implemented for climate change mitigation will positively affect water resources. One example of this is **soil carbon conservation** in agriculture through reduced tillage, erosion prevention and improving the soil's capacity to retain water. More efficient use of fertilisers can also improve water quality.

   Thus, water management can have a significant impact on greenhouse gas emissions and mitigation. Some more examples follow below.

2. **Irrigation and mitigation of emissions.**

   The use of solar energy for irrigation is a potential option for reducing the emissions associated with irrigation. Solar irrigation is an increasingly reliable, relatively low-cost, clean-energy solution for agricultural water management in areas with high incident solar radiation. In many rural areas where reliable access to electricity is lacking or diesel fuel is expensive, solar irrigation initiatives can be a way of providing broader access to energy for agriculture and other uses (FAO and GIZ, 2015). (Turral et al., 2011).

3. Rice production systems (agriculture) and mitigation

Agricultural methane emissions account for more than 50 percent of methane emissions from human activities. One-third of these emissions come from flooded rice production (28 to 44 million tonnes of methane per year).

Emissions during the growing season can be reduced by using various water management practices, such as cultivating aerobic rice and, where conditions allow, alternate wetting and drying. Avoiding water saturation when rice is not grown and shortening the duration of continuous flooding during the rice-growing season are effective options for mitigating methane emissions from rice fields.

4. **livestock production, fisheries and aquaculture and mitigation**

   In India, irrigated pastures are an important part of livestock production systems. As the demand for animal feed increases, their importance is growing. Better pasture management, combined with the use of feed additives that suppress methane fermentation in ruminants, can substantially reduce livestock methane emissions (Turral et al., 2011).

   In inland fisheries and aquaculture, the restoration or creation of riparian habitats can absorb carbon and create suitable environments for capture fish production. The modernisation of fishing and aquaculture facilities also has the potential to contribute towards low-impact fuel-efficient (LIFE) production systems.

5.15 Adaptation Needed in India on Water Impacts

In India, the major rivers get their water flow from the snow and glaciers, and melt in the seasons before the summer monsoon. If this trend of temperature rise continues, the river flow in Himalayan basins may initially increase, but when the reservoirs of snow and ice continue to deplete, the variability of downstream run-off will increase and water flow will subsequently reduce. This will seriously affect the livelihood of people in the river basin including fisheries and agriculture (Grail Research, 2009; Wick & Wittgren, 2009). To address many such water-related threats in India, coping mechanisms in the form of adaptation strategies are urgently needed.

In India, climate change impacts have been one of the main causes of water hazards as felt over last two decades (Water the India Story, 2009). Decreasing glaciers, sudden, unexpected floods, intense heatwaves, and long and prolonged droughts adversely impact both water quality and quantity, leading to huge scarcity of water, spread of disease through both water use and water washed, mass migration in search of water, decrease in irrigation and crop production, decrease in volume of available water for use in industry and in availability of clean drinking water and for basic human hygiene (Brookes et al., 2010; Halady & Rao, 2010). Because of this, India is currently viewed as being "water-stressed". Scientists warn that by 2050, India is likely to slip down the ladder to become a "water-scarce" country (Jain, 2012; Thakkar, 2010; Bhooj, 2009).

The adaptation initiatives which are expected to help in water-related impacts of climate change have been in existence in many different sectors in India. For instance, in **agriculture, irrigation and drainage systems** have been enhanced to bring in more water when needed and drain out excess water in flooding situations.

Also, **Integrated Watershed Management** has been introduced to improve agricultural productivity as well as quality. Exhaustive rainwater harvesting has been introduced in rain-fed regions; the components include Grade line Bund, Drainage Line Treatment, Ground Water recharge, etc. At the same time, micro-irrigation techniques have been applied to improve water-use efficiency. More resilient seeds, to withstand floods and droughts, have also been introduced. (Rao, 2020)

In forest sectors, again irrigation facilities have been improved and suitable canals have been constructed from large rivers (Deshingkar et al., 2008).

In the areas of flood control caused by rivers and water bodies, **many infrastructure initiatives have been constructed in the form of embankments, dams, barrages**, etc. (Jain, 2012). In the coastal zone, multiple options have been considered to overcome water-related problems. These options can protect, adapt, or relocate. We can build new projects and "climate-proof" the existing infrastructure.

Some of the areas in India where adaptation will be most appropriate, either by the public sector or by the private sector, are outlined below.

5.16 Water-related adaptation is required for the following reasons:

- Due to warming temperatures as a result of climate change, there will be higher water demands from the population,
- Due to a change in the hydrological cycle, there will be too much rain in some months (resulting in floods) and too little rain in others (resulting in droughts),
- In India, more areas and hence more population will fall under severe water stress,
- Crop production will suffer due to water shortage and increasing temperature,
- Shortage in water supply will affect power generation and other industries that need large volumes of water,
- There would be increased disease and heat stress due to water shortage,
- Mass migration in search of water will become commonplace,
- Irrigation, consuming 70% of total water needs, will suffer drastically,
- Water-washed diseases will increase,
- Water-related diseases will increase,
- Natural habitats of water-borne diseases will likely expand,
- Brackish water will affect aquaculture.

In India, already, many private sector companies have come forward to voluntarily take up adaptation in water-related impacts (CII 2009). These have been in **rainwater harvesting, community/rural drinking water, de-fluoridisation, salinity mitigation for daily water use,** etc. All the same, much more involvement of the private sector will be welcome and the purpose of this research is to campaign, promote, and motivate other industries to replicate these efforts and renovate and innovate the present practices to take them to the next level of perfection (CII 2009).

5.17 Possible Adaptation Strategies for Industry in India

Private sector options to climate change adaptation relating to water.

Adaptation to Reduce Water Demand in the Face of Shortage

In the face of water shortage, the private sector can help in:

- Recycling and treating industrial wastewater,
- Reducing water demand,
- Incorporating efficient water usage practices,
- Promoting conservation of water,
- Practicing rainwater harvesting
- Investing in wastewater treatment plants to reduce water demand,
- Incorporating 3R: reduce, reuse, recycle,
- Incorporating sewage water management and recycling,
- Monitoring of water consumption,
- Promoting optimisation of operations such as cooling tower operations.

Adaptation to Help Communities Have Access to Water

Industry can help in:

- Providing irrigation water and increasing agricultural production,
- Incorporating community-based watershed development,
- Setting up drinking water and purification systems,
- Developing community water management to reduce water demand,
- Promoting water conservation equipment,
- Setting up used water cleaning and recirculation systems,
- Incorporating a programme for thermal stations to reduce usage,
- Helping farmers halt watershed erosion and grow more food,
- Improving infrastructure and sanitation facilities in neighbouring residential areas,

- Improving sanitation and the availability of drinking water in communities,
- Promoting wastewater treatment plants for groups of small companies.

These initiatives and many more aim to shift India from reactive water management to proactive, data-driven sustainability.

CHAPTER 6

IMPACT ON ECOSYSTEM SECTOR

6.1 Describing Ecosystems

An ecosystem includes all the living things (plants, animals and organisms) in a given area, living with each other, interacting with each other, and with their non-living environments (weather, earth, sun, soil, climate, atmosphere).

This sector is sometimes understood to include also industry, settlements, and communities. The climate change impacts are felt in the ecosystems and concerned industries, settlements, and communities. The most affected ecosystems are usually located in coastal areas and river & flood plains. Wherever the ecosystems are closely linked with climate-sensitive resources and to locations already prone to extreme weather events, the impacts are felt very harshly. In addition, areas undergoing rapid urbanisation are also vulnerable to climate impacts. When severe weather events become more intense or more frequent, the associated ecosystems are severely affected. For instance, when extreme rainfall happens, causing flooding of low- lying areas, industry, settlements and communities are all severely impacted, with the resulting stress rising to catastrophic levels.

More generally, however, an ecosystem is often looked upon as comprising the animals, plants, and microorganisms that live in one place, as well as the environmental conditions that support them. This understanding of ecosystems is thus a community of living things, including plants, animals, and microorganisms, that interact with each other and the physical world. People depend on ecosystems for many benefits, such as food, water, clean air, building materials, and recreation.

Climate change affects ecosystems in many ways. Climate controls how plants grow, how animals behave, which organisms thrive, and how they all interact with the physical environment. As habitats experience different temperatures, precipitation patterns, and other changes, the organisms that make up ecosystems feel the effects. All ecosystems in different regions are experiencing the impacts of climate change, but impacts vary by area and ecosystem.

Associated with ecosystems is the concept of Ecosystem Services, which comprises the products and services provided by ecosystems, such as food, fuel, timber, water, clean air, and medicines. It also includes the regulation of local climate conditions and aesthetic value or cultural identity.

An **ecological threshold** is the point at which there is an abrupt change in an ecosystem quality, property, or phenomenon, or where small changes in one or more external conditions produce large and persistent responses in an ecosystem.

The terms often considered as part of ecosystems are:

- **biome,** which is a large, naturally occurring community of plants and animals occurring in a regional or global land area.
- A **food web,** which is a group of predators and prey that interact in a habitat or ecosystem.
- A **stressor,** which is a factor that reduces the health or productivity of an ecosystem (i.e., causes stress).

6.2 Climate Change and Ecosystem

Ecosystems play a crucial role in both absorbing greenhouse gases and regulating water cycles, acting as natural buffers against climate change and water-related hazards. They absorb greenhouse gases, particularly carbon dioxide, through various mechanisms, and they also play a vital role in water storage, purification, and regulating water flow.

Absorption of Greenhouse Gases:

- Carbon Sinks:

  Natural carbon sinks, such as forests, wetlands, and the ocean, absorb and store a significant portion of the excess carbon dioxide emitted by human activities.

- Mangroves and Seagrasses:

  Wetlands like mangroves and seagrasses are particularly effective carbon sinks, absorbing and storing large amounts of CO_2, helping to reduce greenhouse gas emissions.

- Terrestrial Ecosystems:

  A land-based place where living organisms (biotic) interact with non-living elements (abiotic) like soil, climate, and light, encompassing major biomes such as forests, grasslands, deserts, and tundra, all defined by unique temperature, rainfall, and soil conditions. These crucial ecosystems provide essential resources and support biodiversity, playing a vital role in Earth's overall health.

  Terrestrial ecosystems, including forests, also absorb CO_2 from the atmosphere.

- Ocean's Role:

  The ocean acts as a major carbon sink, absorbing a significant portion of CO_2 from the atmosphere and also absorbing excess heat and energy released by rising greenhouse gas emissions.

How Ecosystems Help Address Climate Change

- Wetlands as Buffers:

  Ecosystems such as wetlands, including mangroves, seagrasses, marshes, and swamps, serve as natural buffers against storm surges, absorbing excess water and precipitation.

- Water Storage and Purification:

  Wetlands provide water storage and purification through their plants and microorganisms.

- Flood and Drought Mitigation:

  Healthy freshwater ecosystems, including wetlands and glaciers, help mitigate the effects of floods and droughts.

- Coastal Protection:

  Coastal ecosystems, like mangroves, protect land from erosion and inundation.

- Climate-smart Agriculture: Climate-smart agricultural practices, such as drip irrigation, can help reduce water demand and conserve freshwater supplies.

Thus ecosystems are extremely crucial to us, but climate change is threatening their existence.

Summarising the Effects of Climate Change on the Ecosystem

1. A significant decline in the population of cold adapted species
2. Coastal flooding and erosion of the coastline
3. Heat- related health problems
4. Increase in greenhouse gas emissions
5. Droughts are common
6. Precipitation pattern shift
7. Storms become more severe
8. Crop losses are astronomical
9. Heat related health problems
10. Climate change significantly impacts ecosystems by raising sea levels, caused by the melting of land-based ice (glaciers and ice sheets) and the thermal expansion of seawater as it warms.

6.3 Ecosystems in India

In India, its diverse geography and climate lead to a variety of ecosystems, including deserts, mountains, forests, wetlands, and coastal areas, each supporting unique biodiversity. The country is recognised as a mega-biodiversity hotspot, home to three distinct ecozones: the Western Ghats, the Himalayas, and the Indo-Burma region.

Here's a more detailed look at some key Indian ecosystems:

6.3.1 Terrestrial Ecosystems:

- **Deserts:**

The Thar Desert, a significant part of India's arid landscape, is home to unique flora and fauna adapted to extreme conditions.

- **Mountains:**

  The Himalayas, a major mountain range, are characterised by diverse ecosystems ranging from temperate to alpine.

- **Forests:**

  India boasts various forest types, including tropical rainforests, deciduous forests, and semi-evergreen forests, each with distinct characteristics and biodiversity.

- **Grasslands:**

  Grassland ecosystems, including steppes and savannas, are found across various regions and support diverse wildlife.

6.3.2. Aquatic Ecosystems:

- **Wetlands:**

  India has a wide array of wetland ecosystems, from high-altitude cold desert wetlands to coastal wetlands, supporting diverse flora and fauna.

- **Rivers:**

  The major rivers of India and their tributaries traverse through diverse geoclimatic zones, displaying high biodiversity.

- **Coastal Areas:**

  India's coastline is home to various coastal ecosystems, including mangroves, estuaries, lagoons, and coral reefs.

Indian Ecosystems:

- **Western Ghats:** A biodiversity hotspot known for its unique flora and fauna.
- **Himalayas:** A major mountain range with a high level of endemism.
- **Indo-Burma Region:** A biodiversity hotspot sharing borders with India.
- **Andaman and Nicobar Islands:** A tropical archipelago with unique flora and fauna.
- **Indo-Gangetic Plain:** A fertile plain with a significant portion of India's biological diversity.

6.4 Ecosystem Services

Ecosystem services in India encompass a wide range of benefits provided by natural ecosystems to humans, including provisioning services like food and water, regulating services like climate regulation and pollination, supporting services like nutrient cycling, and cultural services like recreation and tourism. These services are crucial for human well-being and livelihoods, and their value has been estimated to be a significant portion of India's GDP. Studies have identified various ecosystem services associated with different land commons in India, and the value of these services is often higher in areas with diverse ecosystems, such as the Western Ghats.

Provisioning services consist of all "the products obtained from ecosystems". The following services are also known as *ecosystem goods*:

- food (including seafood and game), crops, wild foods, and spices
- raw materials (including lumber, skins, fuelwood, organic matter, fodder, and fertiliser)
- genetic resources (including crop improvement genes and health care)
- biogenic minerals
- medicinal resources (including pharmaceuticals, chemical models, and test and assay organisms)
- energy (hydropower, biomass fuels)
- ornamental resources (including fashion, handicrafts, jewellery, pets, worship, decoration, and souvenirs like furs, feathers, ivory, orchids, butterflies, aquarium fish, shells, etc.)

Key Ecosystem Services in India:

- **Provisioning Services:**

  These include direct benefits like food, water, timber, fuel, and medicinal resources, as well as raw materials for various industries.

- **Regulating Services:**

  These services help regulate environmental processes, such as climate regulation, air purification, soil erosion prevention, water purification, and flood control.

- **Supporting Services:**

  These are essential for the functioning of ecosystems, including soil formation, nutrient cycling, pollination, and habitat provision for species.

- **Cultural Services:**

  These encompass the non-material benefits humans derive from nature, such as recreation, tourism, aesthetic appreciation, spiritual and religious values, and cultural heritage.

Examples of Ecosystem Services in India:

- **Water Regulation:**

  The Himalayan glaciers and rivers provide a significant source of freshwater for agriculture, industry, and domestic use, as well as supporting biodiversity.

- **Climate Regulation:**

  Forests and other vegetation play a crucial role in regulating the climate, including carbon sequestration and temperature moderation.

- **Pollination:**

  Bees and other pollinators are essential for the reproduction of many plant species, including crops, contributing to food security.

- **Soil Fertility:**

  Ecosystems, including forests and wetlands, help maintain soil health and fertility, which is crucial for agricultural productivity.

- **Biodiversity:**

  India's diverse ecosystems support a wide variety of plant and animal species, contributing to global biodiversity.

6.4.1 Importance of Ecosystem Services:

- **Human Well-being:**

  Ecosystem services are essential for human health, livelihoods, and overall well-being.

- **Economic Benefits:**

  Ecosystem services contribute significantly to India's economy through various sectors, including agriculture, food supply, tourism, and fisheries.

- **Environmental Sustainability:**

  Maintaining ecosystem services is crucial for ensuring the long-term sustainability of India's environment.

6.4.2 Challenges and Opportunities:

- **Decline of Ecosystem Services:**

  Land degradation, deforestation, pollution, and climate change are posing threats to the sustainability of ecosystem services in India.

- **Integrating Ecosystem Services into Planning:**

  Incorporating ecosystem services into urban and regional planning is crucial for ensuring sustainable development.

- **Payments for Ecosystem Services (PES):**

  PES mechanisms can incentivise land owners and communities to conserve and enhance ecosystem services.

- **Raising Awareness:**

  Raising public awareness about the value of ecosystem services and the importance of their conservation is crucial for promoting sustainable practices.

Integrating Capabilities and Ecosystem Services Approaches...

Integrating Ecosystem Services (ES) and Capability Approaches in India provides a comprehensive framework to evaluate how nature, specifically in biodiversity hotspots like the Western Ghats, directly enables indigenous and local community well-being, freedom, and livelihoods. This approach links ecological health with human rights, bridging gaps in conventional conservation policies.

6.4.3 Valuing Ecosystem Services Provided by Land Commons in India

Land commons in India, covering 66.5 million hectares, provide critical ecosystem services valued at approximately USD 90.5 billion annually, crucial for rural livelihoods and ecological stability. These services—spanning forests, pastures, and wastelands—often exceed agricultural value per hectare, countering "wasteland" narratives, though they face decline.

6.5 Climate Change Impacts on Ecosystem

6.5.1 Changes in the Timing of Seasonal Life Cycle Events

Climate is an important environmental influence on ecosystems. Changing climate affects ecosystems in a variety of ways. As the climate changes, some species will adapt by changing their behaviour, physical characteristics, or how their bodies function. Others will not be able to adapt. As a result, climate change could lead to expansions, reductions, or extinctions of some populations. These changes, in turn, can affect the overall biodiversity of a region.

For many species in the ecosystem, the climate where they live or spend part of the year influences key stages of their annual life cycle, such as migration, blooming, and reproduction. As winters become shorter and milder, the timing of these events changes in some parts of the country. Also, warming may force animal species such as birds to migrate to higher latitudes or higher elevations where temperatures are more conducive to their survival. Similarly, as sea level rises, saltwater intrusion into a freshwater system may force some key water species to relocate or die, thus removing predators or prey that are critical in the existing food chain.

Because species differ in their ability to adjust, asynchronies can develop, increasing species and ecosystem vulnerability. These asynchronies can include mismatches in the timing of migration, breeding, pest avoidance, and food availability. Growth and survival are reduced when migrants arrive at a location before or after food sources are present.

6.5.2 Changes in the Timing of Natural Events and Cycles

Many plants and animals rely on cues in nature, including temperature and water conditions, to trigger certain stages of their life cycles. As the climate changes, these cues can change at different rates, or even not at all. As a result, species that depend on one another at certain times of the year may no longer be in sync.

For example, plankton are an important food source for young fish, but they tend to react more quickly to changes in temperature than the fish. This means the plankton might not be as available when growing fish need it most. Also, if a bird migrates at the same time each year, it could reach its destination to find that, due to shifting temperatures, its main food source has grown too early and is no longer available.

Climate change not only affects ecosystems and species directly, but it also interacts with other human stressors, such as development. Although some stressors cause only minor impacts when acting alone, their cumulative impact may lead to dramatic ecological changes. For instance, climate change may exacerbate the stress that land development places on fragile coastal areas. Additionally, recently logged forested areas may become vulnerable to erosion if climate change leads to increases in heavy rainstorms.

6.5.3 Effects of Climate Change on Range Shifts

Climate change affects ecosystems at multiple levels, from the populations that make up ecosystems to the services they provide to communities, economies, and people.

Plants and animals may also change the geographic range they inhabit in response to changing climatic conditions. Changing temperature and water conditions have already altered the ranges of many plants and animals. As temperatures have warmed, some land animals have moved to the typically cooler places.

As temperatures increase, the habitat ranges of many species are moving north and to higher elevations. In recent decades, in both land and aquatic environments, plants and animals have moved to higher elevations. While this means a range expansion for some species, for others it means movement into less hospitable habitats, increased competition, or range reduction, with some species having nowhere to go because they are already at the top of a mountain or at the northern limit of land suitable for their habitat. These factors lead to local extinctions of both plants and animals in some areas. As a result, the ranges of vegetative biomes are projected to change across 5-20% of the land in the United States by 2100.

As rivers and streams warm, warmwater fish are expanding into areas previously inhabited by cold-water species. Range shifts disturb the current state of the ecosystem and can limit opportunities for fishing and hunting.

6.5.4 Food Web Disruptions

The impact of climate change on a particular species can ripple through a food web and affect a wide range of other organisms. Not only is the decline of sea ice impairing polar bear populations by reducing the extent of their primary habitat, but it is also negatively impacting them via food web effects. Declines in the duration and extent of sea ice in the Arctic lead to declines in the abundance of ice algae, which thrive in nutrient-rich pockets in the ice. These algae are eaten by zooplankton, which are in turn eaten by Arctic cod, an important food source for many marine mammals, including seals. Seals are eaten by polar bears. Hence, declines in ice algae can contribute to declines in polar bear populations. The Arctic food web is complex. The loss of sea ice can ultimately affect the entire food web, from algae and plankton to fish to mammals. Source: NOAA (2011)

https://19january2017snapshot.epa.gov/climate-impacts/climate-impacts-ecosystems_.html

6.5.5 Changes in Ecosystem Interactions

Climate change is also changing the way species and populations interact with the environment and one another. These impacts can be felt throughout an entire ecosystem.

For example, climate change is increasing the spread of invasive species in some areas. An invasive species is one that is not native to an area. Invasive species can outcompete native plants and animals, bring in new diseases, and cause other problems. These changes can create significant environmental and economic harm.

As ocean water warms, invasive fish species, such as the tropical lionfish, are expected to move north along the Atlantic coast, threatening native species. This can also hurt humans, as lionfish are venomous and can sting people.

Climate change can also affect food webs. A food web is the whole set of feeding relationships among different organisms in an ecosystem. At the bottom of a food web are organisms like plants and plankton. Other animals, higher in the web, rely on them as food sources. Climate impacts on any part of a food web can affect the whole system, and even other ecosystems altogether. From the example above, if young fish cannot find enough food in the estuaries, their predators in the ocean could feel the effects as well.

6.5.6 Climate Change Affects Ecosystem Services

Ecosystems and their benefits—often referred to as ecosystem services—support key aspects of human existence. As a result, they are the foundation of large parts of the economy. Globally, ecosystem services are worth an estimated $125 to $145 trillion per year.

Climate change is affecting some of the critical services that ecosystems provide to society. For example, ecosystems provide a bounty of food to people. Climate changes, like drought and heat, could affect the availability and quality of some foods, as well as farmers' ability to grow certain crops.

Climate change can also affect ecosystem services such as carbon capture and storage. Forest ecosystems play a critical role in the carbon cycle, helping to absorb carbon dioxide from the atmosphere and store it in roots, soil, and the forest floor. But climate-driven increases in wildfires, flooding, pests, and diseases can limit the ability of an ecosystem to provide this important service.

Climate change impacts affect the livelihoods of millions of people, including fishers, loggers, ranchers, and farmers. For example, the stress caused by rising water temperatures and ocean acidification could cost the shellfish industry hundreds of millions of dollars. Shifting ranges for certain fish mean that fishers may need to travel farther to catch them or purchase new equipment to reach those areas. These shifts are projected to result in losses of hundreds of millions of dollars each year by the year 2100.

For example, in the USA, agriculture will experience economic impacts, though they could be both negative and positive. Climate change threatens to bring increased droughts, pests, and extreme weather events, which would take an economic toll. However, earlier springs and milder winter weather can prolong growing seasons in many regions, increasing crop yields and revenues.

Climate change also affects the tourism and recreation industries. For example, harmful algal blooms already cost the USA nearly $1 billion each year from the loss of recreational fishing and boating revenue. Damage to coral reefs due to climate change is projected to result in $140 billion in total lost revenues to the US recreation industry by 2100. Communities that depend on tourism will feel these economic impacts the most.

Some communities depend on ecosystems and their services more than others, including some indigenous and rural communities. As ecosystems change, some groups may be able to adapt better than others due to geographic, social, or economic factors.

Indigenous communities have a unique relationship with the environment. Climate change impacts on ecosystems can harm indigenous peoples' ability to preserve and sustain their cultural heritage. Shifting animal ranges can disrupt indigenous peoples' ability to hunt and fish as the animals move beyond their tribal lands.

Many rural communities depend on ecosystem-related livelihoods, including farming, fishing, and raising livestock. People in these areas are more likely to see economic impacts when faced with ecosystem challenges, such as invasive species and decreased soil quality.

To preserve the ecosystems from the harmful effects of climate change, Adaptation and Mitigation strategies can be applied. These strategies help ecosystem survival and, in the process, the use of biodiversity and ecosystem services form a part of an overall adaptation strategy to help people and communities adapt to the negative effects of climate change at local, national, regional and global levels.

6.6 Ecosystem in Mitigation and Adaptation to Climate Change

Ecosystems can also assist in the mitigation of, and adaptation to, climate change. The focus is now to examine (i) the threats that climate change poses to ecosystems, (ii) the opportunities to enhance ecosystem resilience to climate change, and (iii) the consideration of how ecosystems and ecosystem restoration can assist climate change mitigation and adaptation mechanisms. The potential and limits of such nature-based solutions to climate change need to be explored and quantified.

6.6.1 The Ecosystem-based Adaptation (EbA) Approach

This approach enables addressing climate change by empowering rural communities to sustainably restore and nurture their ecosystems through nature-based solutions.

https://wotr.org/2022/02/23/what-is-ecosystem-based-adaptation-and-how-can-it-help-india-fight-climate-change/

How Does the Ecosystem-based Adaptation Approach Work?

1. It focuses on restoring, maintaining and improving ecosystems so that they are resilient to climate change. EbA looks beyond the role of ecosystems as being providers of fixed natural resources and instead sees them as a network of interconnected beings that not only generate products such as food but also provide regulating services that affect climate and water cycles, cultural services that provide educational and community benefits and supporting services such as nutrient cycling.

   For example, in watershed development projects, one not only looks at conserving soil and water but also pays attention to the needs of humans, flora and fauna while promoting local biodiversity and maintaining a People's Biodiversity register.

2. It helps people to adapt to the current and future effects of climate change. Healthy ecosystems contribute to the resilience of communities by reducing dependency on environmental resources, identifying new income opportunities, increasing awareness of local biodiversity and ecosystem services and enhancing social and cultural capacities.

   For example, alternative farm-allied and non-farm income sources that are more climate- resilient (apiaries, value addition, oilseed processing, ecotourism) were identified to diversify livelihood opportunities. To state an example, in Purushwadi, ecotourism provides jobs to the youth and income to families from catering.

   (Purushwadi, a small village in Maharashtra's Ahmednagar district, is a renowned ecotourism destination famous for its spectacular firefly display during May and June. Millions of fireflies light up the forest, creating a "magical," tree-illuminating spectacle. It offers a rustic, rural, and digital-detox experience with homestays and camping options for visitors.)

3. It enables participatory governance by calling for collaboration between multiple sectors, strengthening local governance and creating linkages between practice and policy. Traditional knowledge, ongoing research and studies help all stakeholders to make informed decisions that minimise risks and maximise benefits between conservation and development goals.

For example, the Water Stewardship Initiative (2015-2017) implemented as part of a watershed development project in Maharashtra was a collaboration with Hindustan Unilever Foundation. It aimed at building community capacity in the face of droughts through water budgeting, which is basically crop selection based on available water (rainfall and groundwater). This selection is made with the help of local committees after prioritising domestic and livestock drinking water needs and promotion of water-saving techniques.

Ecosystem, mitigation and adaptation

(https://www.cifor-icraf.org/publications/pdf_files/Books/BLocatelli160138.pdf)

6.7 Ecosystems and Climate Change Mitigation

Ecosystems mitigate climate change by removing the greenhouse gas carbon dioxide from the atmosphere. In recent years, roughly a quarter of the carbon dioxide emitted by human activities (due largely to fossil fuel burning and land use change) has been taken up and stored in the plants and soils of terrestrial ecosystems. In this way, ecosystems act as carbon "sinks." As greenhouse gas concentrations continue to rise in the atmosphere, some researchers are exploring ways to manage ecosystems to enhance these carbon sinks beyond the capacity they have exhibited in the past.

Enhanced mitigation potential of many ecosystem types can be achieved through management practices that promote greater carbon uptake and storage. Recent efforts have explored this potential, along with the costs and risks associated with management across a wide range of ecosystem types, to determine the feasibility of implementation at scales likely to have a measurable impact on climate change.

Ecosystem management alone is not enough to address the challenge of rising carbon emissions, but it can play a significant role. Research on carbon uptake and mitigation suggests that a wide range of ecosystem types have the potential to increase carbon storage and/or avoid greenhouse gas emissions with conservation, restoration, and/or improved land management. Ecosystem management thus includes reforesting areas where trees have been cut down, avoiding the conversion of forests to other land uses, and better management of natural forests. Increased carbon storage can also occur in agricultural systems, where better nutrient management may be the most cost-effective approach, and other practices (such as the addition of biochar) may have the greatest potential to improve carbon uptake and storage. Coastal ecosystems such as seagrass beds, mangrove swamps, and salt marshes also act as significant carbon sinks, and ocean organisms act as a biological pump transferring carbon from surface waters into the deep ocean carbon sink.

6.8 Adaptation in Ecosystems

Ecosystems Buffer and Threshold Effects

Ecosystems can serve as natural buffers from extreme events such as wildfires, flooding, and drought. Climate change and human modification may restrict ecosystems' ability to temper the impacts of extreme conditions, and thus may increase vulnerability to damage. Examples include reefs and barrier islands that protect coastal ecosystems from storm surges, wetland ecosystems that absorb floodwaters, and cyclical wildfires that clear excess forest debris and reduce the risk of dangerously large fires.

6.9 Managers can address climate change's impact on ecosystems by Building Ecosystem Resilience:

- Protect and Restore Wetlands:

  Wetlands play a vital role in flood control and carbon sequestration.

- Preserve Coastal Land:

  Protecting coastal ecosystems from development and erosion helps maintain biodiversity and coastal resilience.

- Manage Water Resources:

  Implement integrated water management strategies to ensure water availability and quality, especially in drought-prone areas.

- Preserve and Restore Habitats:

  Protect and restore natural habitats to enhance biodiversity and ecosystem services.

- Control Invasive Species:

  Implement strategies to prevent the spread of invasive species that can disrupt ecosystems.

- Promote Ecosystem-Based Management:

  Integrate ecological considerations into decision-making to ensure the long-term health and resilience of ecosystems.

6.10 How Corporations Are Helping Ecosystems in India

Several organisations are dedicated to managing and restoring ecosystems under threat of climate change through scientific, community-led, and corporate-supported initiatives. These managers work across diverse areas, including forest restoration, marine conservation, wildlife protection, and water management.

6.10.1 Leading Organisations and Initiatives in India

- Watershed Organisation Trust (WOTR): Focuses on community-led watershed development and ecosystem-based adaptation (EbA) to climate change.
- Foundation for Ecological Security (FES): Specialises in the collective management of land and water resources.
- Ashoka Trust for Research in Ecology and the Environment (ATREE): Conducts research on biodiversity conservation and sustainable development, with a specific focus on Open Natural Ecosystems (ONEs).
- IORA Ecological Solutions: Works on climate change, carbon forestry, and biodiversity conservation, supporting government policies.
- Terracon Ecotech: A private firm specialising in restoring biodiversity and creating Miyawaki forests.
- Centre for Wildlife Studies (CWS): Focuses on reducing human-wildlife conflict and promoting agroforestry.
- Sadhana Forest: Focuses on ecological restoration and sustainable living in Auroville, Tamil Nadu.
- Eco-Watch: Established by Suresh Heblikar to focus on environmental conservation and field studies.

6.10.2 Focus Areas for Ecosystem Management

- Forest Restoration & Afforestation: Organisations like Saytrees Environmental Trust and Acacia Eco specialise in large-scale reforestation and Miyawaki methods.

- Marine & Coastal Conservation: Deepika Sharma (THT) and Sri Chakra Pranav (Fishing Cat Conservancy) manage marine protected areas and mangrove ecosystems.
- Aquatic Ecosystems: Projects in the North-Eastern Himalayan Region, supported by GIZ, focus on managing wetlands and aquatic biodiversity.
- Urban Biodiversity: The Centre for Environmental Management of Degraded Ecosystems (CEMDE), in collaboration with the Delhi Development Authority (DDA), has developed six biodiversity parks in Delhi.

CHAPTER 7

FOOD & AGRICULTURE SECTOR

In Agriculture, one includes crop production, horticulture, livestock, fisheries, forestry, and fodder and milk production.

Climate change induced temperature changes and the frequency of droughts and floods are expected to negatively affect crop production, increasing the number of people at risk from hunger, malnutrition, displacement, and migration. In addition, the heating of ocean waters would affect fish life adversely, resulting in a lack of availability of seafood of all kinds.

Agriculture is very sensitive to weather and climate. It also relies heavily on land, water, and other natural resources that climate affects. While climate changes (such as in temperature, precipitation, and frost timing) could lengthen the growing season or allow different crops to be grown in some regions, they will also make agricultural practices more difficult in others. (Precipitation is simply water falling from the sky as rain, snow, sleet, or hail, forming when atmospheric water vapour condenses and becomes too heavy to stay in clouds, eventually falling back to Earth as part of the water cycle.)

The effects of climate change on agriculture will depend on the rate and severity of the change, as well as the degree to which farmers and ranchers can adapt. Agriculture in many countries already has many practices in place to adapt to a changing climate, including crop rotation and integrated pest management. A good deal of research is also underway to help prepare for a changing climate.

7.1 Top Climate Impacts on Agriculture

Climate change may affect agriculture at both local and regional scales.

Key impacts are described in this section.

1. **Changes in Agricultural Productivity**

   Climate change can make conditions better or worse for growing crops in different regions. For example, changes in temperature, rainfall, and frost-free days are leading to longer growing seasons in many regions. However, a longer growing season can have both positive and negative impacts for planting. In some cases, farmers may be able to plant crops which take a long time to mature, while in other cases, farmers may need to provide more irrigation over a longer, hotter growing season. Air pollution may also damage crops, plants, and forests. To explain this, when plants absorb large amounts of ground-level ozone, which happens because of air pollution, they experience reduced photosynthesis, slower growth, and higher sensitivity to diseases.

   The threat of wildfires can also increase with climate change. Wildfires bring in major risks to farmlands, grasslands, and rangelands. Changing temperature and precipitation rates will also expand the occurrence and range of insects, weeds, and diseases and thus could lead to a greater need for weed and pest control.

   The growth of many crops depends on pollination in different regions. Warmer temperatures and changing precipitation can affect when plants bloom and when bees and butterflies, which serve as pollinators, come out to pollinate the crops. If mismatches occur between when plants flower and when pollinators emerge, pollination would surely decrease.

2. **Impacts to Soil and Water Resources**

   There are many animals with limited mobility, like oysters, which are vulnerable to hypoxia, resulting from limited oxygen during climate change, because they can't move to other areas.

   Climate change is expected to increase the frequency of heavy precipitation in some regions, which can harm crops by eroding soil and depleting soil nutrients. Heavy rains can also increase agricultural run-off into oceans, lakes, and streams. This run-off can harm water quality.

When coupled with warming water temperatures brought on by climate change, run-off can lead to depleted oxygen levels in water bodies. This is known as hypoxia. Hypoxia can kill fish and shellfish. It can also affect their ability to find food and habitat, which in turn could harm the coastal societies and economies that depend on those ecosystems.

Sea level rise and storms also pose threats to coastal agricultural communities. These threats include erosion, agricultural land losses, and saltwater intrusion, which can contaminate water supplies. Climate change is expected to worsen these threats.

3. **Health Challenges to Agricultural Workers and Livestock**

   The workers who work in the field of agriculture face several climate-related health risks. These include exposures to heat and other extreme weather, more pesticide exposure due to expanded pest presence, disease-carrying pests like mosquitoes and ticks, and degraded air quality. Heat and humidity can also affect the health of agricultural workers, along with decreasing the productivity of animals raised for meat, milk, and eggs.

   Climate change could affect food security for many households in many countries. Even households which were food secure earlier can run into lack of food in the years to come. Climate change can also affect food security for some indigenous peoples in regions where climate impacts like sea level rise and more intense storms can affect the production of speciality crops. These crops are often key sources of nutrition and may also have cultural and economic importance.

   There can be many strategies to address impacts on agriculture.

   These can be implemented with support from corporate organisations, as indicated below.

7.2 Strategies to Address Impacts on Agriculture on the Part of Corporate Initiative

One can reduce the impact of climate change on agriculture in many ways, including the following:

- Corporate organisations can encourage Climate-smart farming methods by building awareness in farmers to use climate forecasting tools, plant cover crops, and take other steps to help manage climate-related production threats.
- Corporate organisations can help farmers to plant a pollinator-attracting garden with native species. This is just one of many ways to help pollinators like bees and butterflies thrive.
- Livestock producers can get help in recovering methane, a potent greenhouse gas, from biogas created when manure decomposes.
- Corporate organisations can extend training to farmers in the farmlands to reduce runoffs by strategically applying fertiliser. Also, farm animals can be out of streams, and farmers can take more actions to reduce nutrient-laden run-off.
- Corporate organisations can extend training to farmers to boost crop resistance, adopt research-proven ways to reduce the impacts of climate change on crops and livestock in their ability to withstand, tolerate, or recover from biotic stresses (pests, pathogens) and abiotic stresses (drought, salinity, extreme temperatures), both brought about by climate change. Also, efforts should be taken to reduce pesticide use and improve pollination.
- Another very important strategy would be to prevent food waste. Encourage people to donate nutritious, untouched food to food banks and those in need.

7.3 Corporate organisations can build awareness to prevent food insecurity by:

1. Encouraging food production and distribution companies to invest in food storage systems that can withstand extreme weather events.
2. Encourage food production companies to diversify food sources and try sustainable agricultural production techniques to reduce risk.
3. Enhancing awareness in agriculture that it is very important to adopt water management systems that reduce crop damage from floods or droughts.

4. Encouraging Sustainable farming practices such as incorporating no-till agriculture, agroforestry, and cover crops.
5. The banks could support smallholder farmers with access to credit and other services to ground economic empowerment.
6. Public awareness could be enhanced to address food security challenges caused by climate change.
7. Awareness could be enhanced to increase organic carbon in soil to enable water retention in soil, increasing resilience to drought.
8. Education on food preservation techniques, such as refrigeration and dehydration, could be increased.
9. Very important early warning systems should be developed for extreme weather events to enable food production. Also, adaptation approaches can be brought in by leveraging technologies to use data analytics and insights, predictive AI.
10. Corporate organisations can invest in research and development for climate-resilient food crops.

7.4 How Can Corporate Organisations Enhance Awareness to Help Agriculture Adapt to Climate Change?

Corporate organisations can build awareness and invest in a few key strategies that:

Water should be used more efficiently and effectively, combined with policies to manage demand. More irrigation infrastructure would be a solution if the future water supply proves to be adequate to supply the irrigation systems. In agriculture, the system should include better management of water demand as well as the use of advanced water accounting systems and technologies. These approaches would assess the amount of water available, including soil moisture sensors and satellite measurements. These approaches can facilitate techniques such as alternate wetting and drying of rice paddies, which saves water and reduces methane emissions at the same time.

Agriculture should switch to less water-intensive crops. Rice farmers could switch to crops that require less water, such as maize or legumes. This would also help reduce methane emissions, because rice is a major source of agri-food emissions. But a culture that has been growing and consuming rice for thousands of years may not so easily switch to another, less water-intensive, less emitting crop.

Improve soil health. This involves increasing organic carbon in soil, which helps it better retain water and allows plants to access water more readily, increasing resilience to drought. This initiative also provides more nutrients without requiring as much chemical fertiliser -- which is a major source of emissions. Farmers can restore carbon that has been lost by not tilling soil and by using cover crops, particularly with large roots, in the rotation cycle rather than leaving fields fallow. Such nature-based solutions to environmental challenges could deliver 37% of climate change mitigation. However, getting farmers to adopt these practices will take time, awareness-raising and training. In places where farm plots are small and farmers can't afford to let fields lie fallow or even rotate with leguminous crops, improving soil health could pose a challenge.

Further to Awareness Building, certain adaptation approaches could also help in Agriculture:

7.5 Adaptive Approaches to Help Agriculture and Food

Improved Crop production

Crop production involves crop rotation, which is the agricultural practice of growing different types of crops in a planned sequence on the same land across consecutive seasons. To cope with the challenges of climate change, crop production must adapt (e.g. crop varietal selection, plant breeding, cropping patterns and ecosystem management approaches) and become resilient to changes (frequency and intensity).

Integrated practices

Integrated practices that involve a range of stakeholders will ensure greater efficiency in the use of resources and more sustainable management of natural and human-created processes in the landscape. Integration can greatly reduce the pressure on the natural resources and minimise the need for external inputs (e.g. energy, chemical fertilisers and pesticides).

Livestock

Managing Livestock properly can make a large contribution to climate-smart food supply systems. This approach tries to reduce greenhouse gases along the entire supply chain. This relates to feed management, enteric fermentation and manure management.

Forestry

Forests maintain the climate by acting as critical carbon sinks, absorbing atmospheric CO_2 for photosynthesis to reduce global warming. Trees absorb vast amounts of carbon dioxide from the atmosphere—a primary greenhouse gas—and store it in their biomass and the soil, preventing it from contributing to global warming. Forests enhance agricultural productivity through ecological services and support the livelihoods of people. They offer a safety net during 'hunger seasons', provide fuel for cooking, and protect soil and water resources essential for sustainable farming. Thus, preservation of forests would be of utmost importance even for food security.

Also, climate change jeopardises the delivery of goods and ecosystem services, from forests and trees that are essential to livelihoods and food security, to environmental sustainability and to national development. Thus, forests should be preserved for the supply chain of forest products.

Fisheries and aquaculture

Fisheries and aquaculture provide essential nutrition, support livelihoods and contribute to national development. However, because of climate change, the sector is facing significant challenges in maintaining its crucial contribution to these areas. Increasing global demand for fish and aquatic foods, ocean acidification and climate change only add to these challenges. Thus, there is a tremendous need to protect these sectors.

Land and water management

Land and water management is a key element of CSA, community supported agriculture. More productive and more resilient agriculture requires a major shift in the way land and water are managed to ensure that these resources are used more efficiently. Practices to sustainably manage land and water include a broad range of practices and methods, including soil carbon sequestration and the restoration of peatlands (vital wetland ecosystems characterised by waterlogged conditions that cause partially decomposed plant matter to accumulate as peat over thousands of years).

Proactive drought management

Drought is a complex natural hazard which affects all climates and results in socio-economic impacts, the extent of which varies depending on several factors and conditions. Agriculture is the first and most drought-affected sector. Proactive drought management thus focuses on risk reduction, preparedness, and resilience, rather than reactive emergency aid. Key practices include developing early warning systems, implementing water-efficient technologies (drip irrigation), restoring landscapes for water retention, adopting drought-resistant crops, and utilising policy-driven water allocation.

Energy

Agricultural producers can use energy more efficiently as well as reduce their dependence on non-renewable energy sources. Well-planned agricultural systems can also produce energy sustainably without compromising food security, thus contributing to the transition to climate-smart agriculture. This can only happen if existing examples of energy-smart food systems can be scaled up significantly.

Food loss and waste

Food loss and waste amount to a major loss of resources, including water, land, energy, labour and capital and lead to greenhouse gas emissions, contributing to climate change.

Food loss and waste generate 8–10% of global greenhouse gas (GHG) emissions annually, making it a major contributor to climate change. Food waste produces methane (a potent greenhouse gas) as it decays in landfills. Reducing waste saves resources and lowers emissions.

What can the private sector do to help build food security in the face of climate change?

The private sector can step up support for climate-smart agriculture across the agriculture and food value chains and also implement technological interventions to enhance productivity, improve resilience, and reduce GHG emissions.

The private sector can also set up projects aiming to benefit farmers and pastoralists in different communes through the distribution of improved, drought-tolerant seeds, more efficient irrigation, and expanded use of forestry for farming and conservation agriculture techniques. This way, many hectares can be under more sustainable farming practices.

7.6 Indian Context... Climate Change Impacting Agriculture in India

Like other places all around the world, in India too, climate change hugely threatens food security and agriculture through unpredictable monsoons, rising temperatures, and extreme weather, with studies projecting yield declines of up to 47% for rain-fed rice and 40% for wheat by 2080. The resulting impacts include reduced nutritional quality, increased pest outbreaks, soil degradation, and severely affected rural livelihoods.

7.6.1 Key Impacts of Climate Change on Indian Agriculture

- Reductions in agricultural production: By 2050, without adaptation, rice yields are projected to fall by 20%, wheat by 19.3%, and maize by 18%. Extreme, rapid increases in temperature (2.5 to 4.9°C) could bring down wheat yields by 41–52%.
- Unpredictable monsoon & Water Stress: Intermittent and erratic rainfall patterns and droughts are damaging crops and reducing water availability for irrigation. If a 10% reduction in rainfall happens, as estimated, this will result in a 7-8% decrease in rice yields.

- Extreme Weather Events: Increased frequency of floods, cyclones, and heatwaves may destroy crops and disrupt agricultural supply chains.
- Nutritional Value may reduce: Concentrations of essential nutrients like protein, zinc, and iron in crops such as wheat and rice may reduce.
- Livestock and Fisheries: Climate change will negatively affect livestock productivity and reduce fish production.
- Soil Degradation: Rising temperatures contribute to salinity intrusion in soil and groundwater, reducing fertility.

7.6.2 Climate Change Impact on Food Security and Economy

- Price Volatility will happen on account of climate change: This is because agricultural output may reduce, in staples, and this will cause increased food prices, which in turn will threaten food security and make food unaffordable for most of the population.
- Risk of Livelihood on account of climate change: Agriculture is the main livelihood for a large portion of India's population. Reduced yields will directly increase economic risks for farmers.

7.7 Adaptation and Mitigation Strategies

- Climate-Resilient Agriculture (CRA) is very important to be implemented. This would involve adopting heat-resistant and drought-tolerant crop varieties.
- Efficient Water Management: Implementing better irrigation techniques, such as drip irrigation, to cope with water scarcity.
- Policy Intervention: The Indian government is implementing various initiatives to promote resilient agriculture.
- On the private sector side, several Indian companies and startups are driving climate-smart agriculture by focusing on reducing emissions, enhancing soil health, using AI for resource optimisation, and developing climate-resilient crop varieties. Key players include Fasal, CropIn, Ninjacart, DeHaat, and Dharaksha Ecosolutions,

which address challenges like water management, supply chain efficiency, and stubble burning.

7.8 Some Indian Companies Working on Climate Change in India in Agriculture

- **Agri-Tech for Sustainability & Data:**
    - Fasal (Bengaluru): Uses IoT-based sensors and AI to provide farm-level forecasts and irrigation alerts, helping farmers reduce water usage and adapt to climate risks.
    - CropIn: Offers AI-driven, data-driven farm management solutions to create resilient, sustainable food systems.
    - DeHaat: Provides AI-based farming solutions, improving input efficiency and farm productivity.
    - AgNext Technologies: Uses AI for rapid, precise quality assessment to reduce food loss.
    - Stellapps: Focuses on digitising the dairy supply chain to reduce carbon footprints in dairy farming.
- **Sustainable Farming & Soil Health:**
    - ITC (Climate-Smart Agriculture): Implements widespread initiatives to promote sustainable practices and water stewardship among farmers.
    - BioPrime AgriSolutions: Develops bio-molecules to strengthen crop resilience against climate stress.
    - Oorja Development Solutions: Provides solar-powered irrigation to reduce reliance on diesel pumps.
    - ACSEN Agriscience: Focuses on breeding climate-resilient, high-yielding seeds.
- **Waste Reduction & Alternative Materials:**
    - Dharaksha Ecosolutions: Converts agricultural waste (stubble) into biodegradable packaging, directly addressing stubble burning.

- Ninjacart: Optimises supply chains to reduce post-harvest food waste.

- **Initiatives & Tech-Platforms:**
  - The PRANA Initiative (by The Nature Conservancy): Promotes no-burn agriculture and regenerative practices in Northwest India.
  - Bhoomi Farms: Focuses on regenerative agriculture practices.

These companies are essential in transitioning Indian agriculture toward a more sustainable and climate-resilient future.

Several companies and startups in India are working on climate change mitigation and adaptation within the food and agriculture sector, focusing on sustainable practices, reducing carbon footprints, and increasing climate resilience. Key areas of innovation include precision farming, regenerative agriculture, waste reduction, and solar-powered solutions.

In addition, some other prominent companies working on climate change in food and agriculture in India are listed below.

7.9 Climate-Tech & Precision Agriculture

- Varaha: A climate-tech startup focusing on carbon sequestration, which recently achieved major milestones in carbon credit issuance for regenerative farming.
- Fasal: An AI-powered IoT platform for horticulture that helps farmers save water, reduce pesticide usage, and optimise irrigation to boost climate resilience.
- Cropin: Uses AI and data analytics to promote sustainable, climate-smart agriculture and provides early warning systems for crop diseases.
- Absolute (absolute.ag): Focuses on leveraging technology for sustainable agriculture practices.
- BioPrime AgriSolutions: Develops biological products to help crops adapt to changing climatic conditions and reduce chemical residue.

- Fyllo: Provides AI-driven, data-based insights for precision farming, helping to optimise resources and reduce environmental impact.
- Satyukt Analytics: Uses satellite data to provide SaaS solutions for data-driven, sustainable farming.

7.10 Sustainable Farming & Regenerative Agriculture

- Oorja Development Solutions: A social enterprise that provides solar-powered irrigation and farming-as-a-service, reducing reliance on diesel pumps.
- Eeki Foods: Develops climate-proof growing chambers that use 80% less water than traditional methods, allowing for sustainable, year-round farming.
- Krishi Janani: Focuses on regenerative agriculture, working with smallholder farmers to restore soil health and biodiversity.
- Edible Routes: Promotes urban farming and regenerative, diverse ecosystems to produce food sustainably.
- Bombay Hemp Company (BOHECO): Promotes hemp cultivation as a sustainable alternative in agriculture and textiles.

7.11 Supply Chain & Waste Reduction

- Waycool: A food distribution company that optimises the supply chain to reduce food waste, supporting 35,000+ farmers.
- Ninjacart: Utilises AI to streamline the agricultural supply chain, significantly reducing food wastage.
- S4S Technologies: A solar-powered food processing company that enables small farmers to reduce post-harvest losses and improve sustainability.
- Farmtheory: Focuses on mitigating food loss and boosting farmer income by reducing waste at the source.
- Intello Labs: Uses computer vision and AI to assess the quality of produce, reducing waste in the supply chain.
- Ecozen Solutions: Provides solar-powered cold storage (Ecofrost) and irrigation systems to improve sustainability and reduce food loss.

7.12 Bio-inputs & Soil Health

- Essential Bio Sciences: Focuses on bio-based farming solutions, organic fertilisers, and microbial inoculants to restore soil health.
- Gaiagen Technologies: Specialises in bio-based pest management, including insect pheromones, to reduce reliance on synthetic pesticides.
- ABTEC: Develops microbial-based solutions to enhance soil fertility and crop resilience.
- Varsha Bio Science: Advances agricultural biotechnology, focusing on microbial inoculants for soil regeneration.

Indian companies are combating agricultural climate change through sustainable technology, carbon reduction, and climate-smart farming, notably with startups like Ecozen (solar irrigation) and Eeki Foods (controlled-environment agriculture). Major corporations like ITC implement Climate-Smart Villages, while Agri-tech firms like Unnati and SLCM focus on reducing post-harvest losses and improving soil health to build climate resilience.

These organisations are crucial in transitioning Indian agriculture from high-emission, vulnerable practices to resilient, sustainable, and productive systems.

We know how the agriculture sector in India plays a very crucial role in the country's economic progress. However, climate change brings enormous risks such as less agricultural output, price instability, soil degradation, and reduction in farming cycles.

Various service providers, such as private agribusiness firms and non-governmental organisations (NGOs), must be involved in the transmission of agricultural technologies with information, skills, and expertise, while also leveraging Corporate Social Responsibility (CSR) funds.

Two examples are given below.

1. **ITC Limited:**

   The company contributed to addressing climate change risks by contributing towards improved agricultural practices. The

Sustainable Agriculture programme under CSR attempts to de-risk farmers from erratic weather events through the promotion of climate-smart agriculture premised on the dissemination of relevant packages of practices, adoption of appropriate mechanisation, and provision of institutional services. Under the company, its Agri-Business Centres delivered extension services, arranged Agri-credit linkages and established collective input procurement. Also provided agricultural equipment for hire. In pursuit of the company's long-term sustainability objective of increasing soil organic carbon, a total of 2,801 compost units were constructed during the year.

2. **JSW Steel Limited:**

   JSW Foundation closely works with 6000+ farmers covering 5,800 acres of cultivable land through a participatory approach, striving to align with local, national, and global goals (Sustainable Development Goals) while aiming to increase farm and off-farm incomes. This is achieved through capacity building of the farmers, promoting an improved package of practices, advocating less water-intensive farming techniques, and forging strategic partnerships. The emphasis is on the market-led sustainable enhancement of profits to make Agri-livelihoods a preferred mode of occupation.

Managers and initiatives helping agriculture and food security in India against climate change focus on sustainable practices, technology, and policy interventions. To help India strengthen its climate resilience, the corporate community is working with the government to promote both adaptation and mitigation measures across a broad range of areas under agriculture and food security.

Thus, private sector plays a critical role in climate-resilient agriculture and thereby food security by investing in innovation, sustainable supply chains, and technology to meet climate goals. Key areas include developing drought-resistant seeds, advanced irrigation, and financing climate-smart practices to address a potential $2 trillion adaptation market in the next few years.

CHAPTER 8

IMPACT ON DISEASE & HEALTH SECTOR

8.1 Climate Change and Its Impact on Health

The climate change impacts will act as a major threat to global health, increasing risks through intense heatwaves, extreme weather, floods, storms, fires & droughts, reduced air quality, and food/water insecurity. It may cause cardiovascular and respiratory diseases, vector-borne diseases (e.g., malaria, dengue), and mental health issues.

These changes are predicted to alter the health condition of millions of people through the prevalence of diseases, as mentioned above, induce heat stress, increased deaths, malnutrition, and injury. In addition, increased malnutrition, air-borne disease, diarrheal disease, and malaria in some areas will increase public health vulnerability. As a result, development goals will be threatened by long-term damage to health systems and quality of life.

Climate change is expected to exacerbate health problems that already pose a major burden to vulnerable populations.

For instance,

1. Certain population groups have higher susceptibility to climate-sensitive health impacts owing to their age (children and elderly), gender (particularly pregnant women), social marginalisation (associated in some areas with indigenous populations, poverty or migration status), or other health conditions like HIV. The socio-economic costs of health problems caused by climate change are considerable.
2. Many infectious diseases, like air-borne and water-borne ones, are highly sensitive to climate conditions. There has always been a

correlation between temperature and diarrhoea. A main concern in both developed and developing countries is the increase in the spread of diarrhoeal diseases, as found by various research studies.

3. Climate change expands the transmission season and lengthens the geographical range of many diseases like malaria and dengue. For example, the conditions for dengue transmission are likely to expand significantly across the globe.
4. Climate change will also bring new and emerging health issues, including heatwaves and other extreme events. Heat stress can make working conditions unbearable and increase the risk of cardiovascular, respiratory and renal diseases. Also, it is estimated that 22.5 million people are displaced annually by climate or weather-related disasters, and these figures are expected to increase in the future.
5. Malnutrition and undernutrition have been highlighted as a concern for a number of developing countries in Africa, Asia and Latin America, which involves the impacts of climate change on food security, particularly in relation to floods and drought.

While these risks and challenges are big, there are also many solutions, which constitute the adaptation strategies.

To help scale up adaptation action in countries in the area of human health, various solutions can be arrived at, which would be discussed subsequently.

8.2 Some further specific impacts of climate change on health:

1. **Increased Heat-Related Deaths and Illnesses:**

   Rising temperatures, on account of climate change, often lead to heatstroke, exhaustion, and other heat-related illnesses, particularly among vulnerable populations.

2. **Changes in Infectious Disease Spread:**

   Warmer temperatures can extend the geographic range of disease-causing insects like mosquitoes, increasing the risk of malaria, dengue fever, and other infectious diseases.

3. **Threats to Water and Food Security:**

   Climate change disrupts water resources through drought and floods, and impacts crop yields, often leading to water-borne and foodborne diseases and malnutrition.

4. **Increased Allergies and Asthma:**

   Warmer temperatures and changes in precipitation patterns (rain, snow, hail, etc.) can increase pollen and mould, exacerbating allergies and asthma.

5. **Air Pollution Impacts:**

   Climate change often worsens air quality through increased ground-level ozone and wildfire smoke, leading to respiratory illnesses and other health problems.

6. **Mental Health Impacts:**

   Extreme weather events and climate-related displacement can cause stress, anxiety, and post-traumatic stress disorder.

7. **Disruption of Food Systems:**

   Changes in climate can disrupt agricultural production, leading to food insecurity and increased hunger. This leads to the breakdown in the flow of food from production to consumption, caused by natural disasters due to climate change.

8. **Exposure to Hazardous Chemicals:**

   Extreme weather events, such as floods and other disasters, can contaminate water and soil, increasing exposure to hazardous chemicals and toxins.

9. **Displacement and Migration:**

   Climate change sometimes induces the displacement of communities, which can lead to overcrowding and increased risk of infectious diseases in new areas.

10. **Increased Risk of Chronic Diseases:**

    Climate change impacts can increase existing chronic conditions like heart and respiratory diseases. This is because climate change accelerates chronic diseases like cardiovascular, respiratory, and kidney disorders through intense heatwaves, poor air quality, and wildfire smoke. Rising temperatures and extreme weather increase mortality rates in vulnerable populations (the elderly, low-income). Additionally, shifting climate patterns expand the range of vector-borne illnesses.

Climate change–related weather events have strained healthcare workers, disrupting physical infrastructure and interrupting supply chains, making it more difficult to treat patients and threatening healthcare quality and safety. Shortages of medications and hospital facilities unable to operate in extreme weather are recent examples of climate change's impact on health globally. For example, many tornadoes damage drug warehouses and destroy medications as well as pharmaceutical raw materials, worsening medication shortages for drugs used in surgery and cancer treatment.

Flooding in India (one of the world's largest manufacturing locations of pharmaceuticals) has damaged healthcare infrastructure which has affected patient care around the world. Environmental changes have caused harm to human health at or beyond the scale of the healthcare systems' contribution to patient safety events. All the same, approaches are emerging to address this form of preventable harm to human health.

8.3 Climate Change Impacts on the Healthcare Industry

We know that climate change harms human health... for example, through increasing heat-related stress, acute myocardial events, and cerebrovascular events. Worsening air pollution results in acute exacerbations of asthma and vector-borne diseases. Thus, the **healthcare industry is required to be very active in addressing these impacts on human health**. It has also emerged that, among other alarming facts, heat-related mortality for people above age 65 has increased by more than 50% in just the past 20 years, and a synthesis of systematic reviews showed that globally, climate change has worsened morbidity and mortality related to cardiovascular, respiratory, and infectious diseases.

8.31 Solutions to address the health impacts of climate change:

Addressing climate change impacts on health requires strengthening health systems to handle climate-sensitive diseases, heatwaves, and extreme weather conditions.

Key actions include improving early warning systems, enhancing healthcare facility resilience, boosting public awareness, and reducing emissions in the health sector.

Key Strategies for Mitigation and Adaptation:

- **Policy and Infrastructure:**
  - Action Plan Implementation: Implementing Action Plans on Climate Change to create resilient healthcare systems.
  - Health System Resilience: Upgrading ICUs, laboratories, and blood banks to withstand extreme weather events.
  - Heat Action Plans (HAPs): Implementing heat-index-based early warning systems, urban heat island management, and enhancing worker safety protocols.
  - Sustainable Healthcare: Decarbonising the healthcare supply chain and adopting clean energy in healthcare facilities.
  - ***Disease Surveillance and Health Services:**
    - Climate-Sensitive Disease Control: Strengthening surveillance and response for vector-borne (malaria, dengue) and water-borne (cholera, typhoid) diseases.
    - Emergency Care: Improving disaster preparedness, especially in rural areas, to ensure continuous care.
    - Mental Health Services: Integrating mental health support into disaster management protocols to treat distress from environmental events.
- **Education and Community Action:**
  - Public Awareness: Educating communities on heat risks, safe water practices, and nutrition.
  - Training Professionals: Training health workers to recognise and treat climate-related illnesses on an urgent basis.

(https://pmc.ncbi.nlm.nih.gov/articles/PMC3114809)

However, the healthcare industry itself contributes to climate change, and therefore, indirectly, to harming human health.

8.4 Contribution of the Healthcare Industry to Climate Change

Because of the contribution of Healthcare industry to climate change, preventing healthcare pollution is just as significant as preventing medical errors to reduce patient harm. Unlike medical errors, which have a direct relationship between care provision and an individual patient, the harms from healthcare pollution have indirect effects that are difficult to attribute to any one patient, even as these harms become more difficult to ignore. Governmental healthcare agencies and healthcare systems around the world are becoming increasingly concerned with the impact of climate change on health, as well as the impact of healthcare-related pollution on climate change. These agencies are calling for decarbonisation targets. Evidence-based strategies are required to accelerate healthcare **decarbonisation** and avert the worst predicted harms to health and healthcare systems.

Decarbonising the healthcare industry involves transitioning to 100% clean, renewable electricity, constructing net-zero emissions buildings, and sustainable supply chain management to reduce the sector's significant climate footprint. Key actions include reducing wasteful, low-value care, adopting low-carbon pharmaceutical alternatives, and implementing circular, waste-free operational practices.

8.4.1 Key Strategies for Healthcare Decarbonisation:

- Energy Transition: Using Powering facilities with 100% renewable energy and reducing reliance on fossil fuels, including diesel generators.
- Infrastructure & Operations: Investing in energy-efficient, zero-emission buildings and sustainable, low-carbon transportation for staff and patients.
- Supply Chain & Procurement: Shifting to sustainable, low-carbon, and locally sourced medical supplies and pharmaceuticals, reducing waste, and improving circular economy practices.

- Operational Efficiency: Eliminating unnecessary, low-value clinical practices and diagnostics that generate waste without improving patient outcomes.
- Focus Areas: Targeting emission-intensive areas such as surgical procedures (which generate high solid waste), anaesthetic gases, and transportation.

8.5 How Healthcare Systems Can Mitigate Their Own Contribution to Climate Change

Worldwide, the healthcare industry contributes to about 5% of total greenhouse gas emissions and similar toxic air emissions. These emissions result from healthcare facilities' on-site fuel combustion, emissions from purchased energy, and all other emissions associated with the healthcare supply chain of goods and services. A major part of emissions arises from the supply chain, primarily from pharmaceuticals, chemicals, and medical devices. These emissions could be reduced by limiting causes of emissions, such as inhaled anaesthetics and low-value care.

Inhaled anaesthetics have been found to be one of the most well-known, high-impact, and feasible areas to mitigate emissions. Inhaled anaesthetics are potent greenhouse gases that are released directly from healthcare facilities. Safe, effective alternatives exist for both waste reduction and selection of lower-polluting general anaesthetic medications. Comparing five inhaled anaesthetics, it has been found that two anaesthetics, desflurane and nitrous oxide (N2O), account for much of the greenhouse gas emission related to all inhaled anaesthetics. These findings suggest that hospital systems can immediately improve their greenhouse gas pollution by removing desflurane from hospital formularies in favour of lower-emission alternatives such as sevoflurane. Desflurane is often seen to have clinical superiority over sevoflurane in certain situations; however, professional societies are calling on their members to reconsider this position in the face of environmental impacts and changing climate. Hospitals can also use centrally piped N2O in favour of portable tanks, though recent research shows as much as 70% to 95% of N2O is leaked through central systems prior to use. This needs to be controlled too.

8.6 Corporate Initiatives to address Climate Change impacts on Health.

Climate change impacts health through various pathways, including increased heat-related illnesses, changes in the spread of infectious diseases, and threats to food and water security. Extreme weather events also contribute to mental health problems and displace communities. These effects disproportionately impact vulnerable populations like children, the elderly, and those with pre-existing conditions.

Addressingclimatechangeimpactsonhealthrequires **strengthening health systems to handle climate-sensitive diseases, heatwaves, and extreme weather conditions**.

Key actions include improving early warning systems, enhancing healthcare facility resilience, boosting public awareness, and reducing emissions in the health sector. Corporate initiatives can help in enhancing these actions.

Corporate initiatives addressing the intersection of health, disease, and climate change focus on two main pillars: mitigation (reducing the healthcare sector's carbon footprint) and adaptation (preparing health systems and communities for climate-related illnesses). Required actions may include investing in developing better healthcare facilities, sustainable, low-carbon healthcare infrastructure, greening healthcare supply chains, and using technologies to manage climate-sensitive diseases like heatstroke and vector-borne illnesses.

These corporate initiatives within the healthcare industry to address climate change can be categorised into operational changes, supply chain improvements, and advocacy efforts. Such actions aim to reduce emissions, increase resilience, and influence policy to create a more sustainable healthcare system.

8.6.1 Operational Changes:

- **Reducing Energy Consumption in Hospitals:**

Implementing energy-efficient practices, such as using LED lighting, optimising HVAC systems, and investing in renewable energy sources (solar, wind) can significantly reduce the carbon footprint of medical establishments.

- **Waste Reduction:**

  Minimising single-use medical supplies through reprocessing and reuse, and adopting waste-to-energy technologies, can also lower the environmental impact.

  Also, implementing programmes to reduce food waste in hospitals and healthcare facilities can significantly reduce emissions.

- **Sustainable Fleet Management**:

  Transitioning to electric or hybrid vehicles for ambulances, shuttle vans, and supply transportation can decrease emissions.

- **Promoting Sustainable Commuting**:

  Encouraging employees to use public transportation, carpool, or telecommute can also help reduce emissions.

- **Green Procurement:**

  Prioritising environmentally friendly products and services in purchasing decisions can further reduce the carbon footprint.

8.6.2 Supply Chain Improvements:

- **Decarbonising the Supply Chain:**

  Collaborating with suppliers to reduce emissions throughout the healthcare supply chain is crucial.

- **Choosing Low-Carbon Medical Products:**

  Prioritising the use of energy-efficient and low-carbon medical devices and equipment.

*** Supply Chain and Product Sustainability**

Reducing "Scope 3" Emissions: Pharmaceutical companies like GSK are reducing emissions from their supply chains. (Scope 3 emissions are not produced by the company itself and are not the result of activities from assets owned or controlled by it, but by those that it's indirectly responsible for up and down its value chain). An example of this is what we buy, use and dispose of products from suppliers.

- **Greener Medical Products**: Switching to lower-carbon metered-dose inhalers (MDIs) and reducing the use of high-impact anaesthetic gases (e.g., reducing desflurane).
- **Sustainable Sourcing**: Companies are implementing "take-back" programmes for unused medicines and using sustainable packaging.

8.6.3 Advocacy and Collaboration:

- Advocating for Policy Changes:

  Working with government and industry leaders to promote policies that support climate action and sustainable healthcare practices.
- Collaborating with Other Sectors:

  Addressing climate change requires cross-sectoral collaboration and partnerships to ensure effective and sustainable solutions.
- Supporting Research and Innovation:

  Funding research on climate-resilient healthcare systems and innovative solutions can help drive progress.
- Green Buildings:

  Some hospitals and healthcare facilities are being designed and built with sustainability in mind, using green building materials and implementing energy-efficient technologies.

8.6.4 Using Green and Climate-Resilient Healthcare Infrastructure

- Solar-Powered Hospitals: Organisations are increasingly installing solar panels and on-site renewable energy to ensure continuous, low-carbon power, especially in developing regions.
- Energy Efficiency Upgrades: Major health systems are investing in energy-efficient heating, cooling, and lighting (LEDs) to reduce emissions. (LED refers to light-emitting diode)
- Waste Reduction and Circular Economy: Companies are reducing single-use plastics in medical settings, using reusable equipment, and managing pharmaceutical waste.

- Water Conservation: Many hospitals are implementing water-saving fixtures and rainwater harvesting to ensure water security during droughts.

8.6.5 Direct Initiatives on Climate-Related Diseases

- Heat-Health Action Plans: Companies are developing protocols to protect outdoor workers and communities, including distributing air conditioners during heatwaves, creating cooling stations, and using smart wearables to monitor body temperature.
- Air Quality Monitoring: The Alliance for Clean Air, supported by companies like IKEA and GSK, is mapping pollution and reducing black carbon emissions, which contribute to respiratory and cardiovascular disease.
- Vector-Borne Disease Control: Pharmaceutical companies are increasing R&D for treatments of diseases that are spreading to new regions due to warmer temperatures, such as malaria and dengue fever.

8.6.6 Technological Innovation and Digital Health

- Telehealth and Virtual Care: Promoting and reducing emissions from patient travel by utilising virtual, on-demand, and remote monitoring services.
- AI for Climate-Health Monitoring: Google is using AI to create "Flood Hub" platforms for real-time flood data and developing predictive tools to help hospitals prepare for health surges during extreme events.
- Smart Wearables: Promoting devices that track environmental conditions and health, such as heatstroke detection sensors.

8.6.7. Corporate Coalitions and Pledges

- The WHO-led ATACH: The Alliance for Transformative Action on Climate and Health works with companies to support the COP26 health commitments for low-carbon, sustainable systems. (COP26 is a major event bringing together leaders from all countries in the world to agree on how to step up global action to solve the climate crisis. Almost 30 years ago, global leaders came together to collectively address climate change for the first time.)

- Race to Zero: Many health institutions and pharmaceutical firms are signing pledges to achieve net-zero emissions by 2050 or earlier.
- Global Health Equity Network (GHEN): Members like Burjeel Holdings are focused on ensuring that climate change adaptation efforts do not worsen existing health inequalities.

These initiatives are increasingly driven by the need to manage risks, improve operational efficiency, and respond to the growing demand for sustainable, healthy communities.

8.7 Specific organisations in India to address climate-related diseases

Corporate initiatives in India to address climate-related diseases are increasingly focused on building climate-resilient health systems, reducing air pollution, and tackling vector-borne illnesses through Corporate Social Responsibility (CSR) and Environmental, Social, and Governance (ESG) frameworks. These efforts align with the National Action Plan on Climate Change & Human Health (NPCCHH).

Key corporate initiatives and sectors involved include:

8.7.1. Healthcare Infrastructure & Resilience

Corporations are investing in "green" hospitals and mobile health services to treat climate-sensitive diseases (heatstroke, respiratory issues, water-borne diseases) in vulnerable areas.

- JSW Foundation: Launched the JSW Centre for Climate Change and Health to address heat impacts, vector-borne diseases, and nutrition. It focuses on decarbonising hospitals and implementing "green hospital" guidelines. JSW Foundation (erstwhile Jindal Education and Medical Trust and Jindal South-West Foundation) is the social development arm of JSW group, governed by the ideology that every life is important and must be given fair opportunities to make the best out of it.

- GAIL (India) Limited: Implemented "Project Aarogya," utilising mobile medical units (MMUs) to provide healthcare in multiple states, investing over ₹25 crore in health, water, and sanitation.
- ICICI Foundation: Strengthened health infrastructure by supplying 500+ dialysis units across 16 states.
- SBI Foundation: Operates "Sanjeevani" mobile clinics that provide doorstep OPD (outdoor services) and medicines in remote, vulnerable areas.

8.7.2. Air Pollution and Respiratory Health

With air pollution posing severe health risks, several companies are implementing initiatives to reduce emissions and improve air quality.

- Wipro: Focuses on reducing emissions through energy-efficient buildings and electric vehicle adoption, targeting net-zero emissions by 2040.
- Cummins India Ltd.: Developed "Oxygen Hubs" in Pune by converting non-forest land into green cover to combat urban pollution.
- Dalmia Cement: Adopted alternative fuels and energy-efficient technologies to reduce pollution.

8.7.3. Water-Borne Disease Prevention

Climate change is exacerbating water scarcity, leading to poor sanitation and diseases.

- ITC Limited: Its "Social & Farm Forestry Programme" and "Watershed Development" efforts have made it water-positive for over 23 years, providing secure water access and reducing sanitation-related health risks.
- Tata Steel: Implements the "Thrive Water Security Program," which focuses on building check dams and rehabilitating water bodies in drought-prone regions.
- Coca-Cola Foundation: The "Anandana" program focuses on rainwater harvesting and water replenishment in rural communities.

8.7.4. Vector-Borne Diseases and "One Health"

Companies are beginning to adopt a "One Health" approach, acknowledging the link between ecosystem health, animal health, and human diseases.

- JSW Foundation: Incorporates One Health strategies to address vector-borne disease outbreaks linked to changing climate patterns.
- Amrutanjan Healthcare: Invests in animal welfare, including vaccinations and healthcare, which contributes to preventing zoonotic disease transmission.

8.7.5. Disaster Resilience

- Reliance Industries & Larsen & Toubro: Improving community readiness and response systems through disaster resilience projects in high-risk areas, mitigating injuries and illnesses from extreme weather.

Future Directions

In the face of climate change impacts on the health of communities vulnerable to such impacts, it is urgent to strengthen the healthcare systems to handle climate-sensitive diseases, heatwaves, and extreme weather conditions.

In its present state, healthcare itself is a significant contributor to carbon footprint, which needs to be reduced. At the same time, it must respond to increasing disease burdens and address medical problems arising from climate change impacts. Hence, the healthcare industry and hospital systems need to work together to mitigate climate change and to conceptualise measures and strategies that hospitals can use to address the increasing health problems caused by climate change. Thus, apart from measures to reduce resource consumption and emissions within their own facilities and their supply chain, there is an urgent need for strategic enhancement of healthcare services to address the health distress of communities.

CHAPTER 9

CASE STUDIES ON HOW THE MANUFACTURING SECTOR AND SERVICE SECTOR ARE ADDRESSING CLIMATE CHANGE

9.1 Case 1: Exploring how semiconductor manufacturing has adapted to address climate change in India.

Semiconductor is the process of creating integrated circuits commonly known as microchips. These tiny silicon-based chips form the backbone of modern electronics. From the device one uses as a mobile phone, to laptop computers, to the smart meters monitoring the home's energy usage, semiconductors are everywhere. With this varied level of usage, the semiconductor industry leaves an enormous carbon footprint behind. The manufacturing process of these chips is complex and requires a high degree of precision, cleanliness, and control. While the world shifts towards the adoption of green tech solutions to tackle climate change by reducing emissions, which requires the use of silicon chips, the semiconductor manufacturing itself leaves huge carbon footprints.

9.1.1 Semiconductor Fabrication Unit (FAB)

A semiconductor fabrication unit, also known as FAB, is the manufacturing plant where the raw silicon wafers are turned into integrated chips or ICs. Semiconductor manufacturing involves the use of various raw materials such as silicon, copper, aluminium, and a range of rare elements. Additionally, the process also requires substantial amounts of water and energy. According to one estimate, fabricating semiconductor circuits on a 150-mm wafer uses 285 kWh of power or 1.6 kWh per square centimetre. Moreover, it contributes to 31% of the global greenhouse gas emissions. This is higher than what electricity generation contributes, which is 27%. The consumption of these raw materials not only depletes natural resources but also leads to a considerable amount of waste. Furthermore, the extraction and processing of these materials contribute to greenhouse gas emissions.

Water plays an indispensable role in the manufacturing process of semiconductors. The plant requires a huge amount of water supply to meet the great upsurge in demand. This is because a substantial quantity of water is required in the process. The manufacturing of these chips is a very energy-intensive process too. It also generates a massive amount of waste in production.

9.1.2 Exploitation of Water at a Semiconductor FAB

With the huge shortage of semiconductors in the market, manufacturers are facing increasing pressure to increase production. However, at the same time, they face the pressure to adopt sustainable and efficient practices in their operations.

Water serves as a major component here and has a chief role in silicon chip production. It is used to rinse and clean the silicon wafers during the manufacturing process. Also, the water that is used is not the regular one but the 'Ultrapure' water, which is way purer than the drinking water. In order to make 1000 gallons of Ultrapure water (UPW) or highly purified water (HPW), it takes around 1500 gallons of municipal water, and the semiconductor fabs use 5 million gallons daily.

9.1.3 Initiatives to address climate change and Decarbonisation Targets:

Semiconductor manufacturing, with its high energy consumption and substantial greenhouse gas emissions, poses a challenge to the sustainable development and decarbonisation goals. To address this, these industries are trying to focus on improving efficiency, adopting cleaner technologies, and developing green initiatives. This industry is very resource-intensive and poses ecological issues.

These are some of the measures that can be taken to improve the situation and promote sustainability:

9.1.4 Corporate Initiative to Shift to Renewable Energy:

A lot of fabs use energy driven by fossil fuel, both in the manufacturing process and in the power generation segment of the plant. Switching to renewable energy sources such as solar and wind helps mitigate carbon emissions.

To satisfy the need for technology and electrical devices in our daily lives, the demand for semiconductors will only continue to rise. Although the production of semiconductors is essential to the current digital world, it also has a big effect on climate change. The industry has a large environmental impact, from consuming energy and raw materials to adding to global greenhouse gas emissions. However, its influence can be reduced with deliberate effort, legislative modifications, technological improvements, and sustainable practices. As time goes on, it becomes more crucial than ever for us to strike a balance between the needs of technological development and the pressing need to safeguard and maintain our environment.

9.1.5 Addressing Climate Change Across the Entire Supply Chain

The GHG Protocol Corporate Standard categorises greenhouse gas emissions associated with a company's Corporate Carbon Footprint (CCF) as Scope 1, Scope 2, and Scope 3 emissions. However, this categorisation does not include the Product Carbon footprint (PCF), which is the total amount of greenhouse gas emissions generated by a product or a service over the different stages of its life cycle.

1. Scope 1 emission includes direct emissions from the company's owned or controlled sources. It encompasses all energy uses as well as process emissions that are released from burning PFCs (perfluorocarbons, PFCs). PFCs are potent, human-made greenhouse gases with extremely high global warming potential (GWP) and atmospheric lifetimes spanning thousands of years.

   PFCs are used in semiconductor manufacturing, aluminium smelting, and refrigeration. They act as alternatives to ozone-depleting substances. Though emissions have reduced in some areas, with much corporate effort, they remain potent climate drivers. They are highly stable greenhouse gases used in semiconductor manufacturing, often requiring high-temperature catalytic destruction (often referred to as "burning" or abatement) at temperatures exceeding 1000 degrees centigrade. They need to decompose, as they are resistant to the natural degradation of chemicals and other gases during the manufacturing process.

2. Scope 2 emissions include indirect emissions from purchased energy, such as electricity used by the company.
3. Scope 3 emissions include indirect value chain emissions, which include the upstream and downstream emissions. Upstream emissions encompass the indirect greenhouse gas emissions within a company's value chain related to purchased or acquired goods and raw materials, while downstream emissions include the indirect greenhouse gas emissions within a company's value chain related to sold goods and services.

Thus, semiconductor companies have begun to measure and take steps to minimise Scope 1 and Scope 2 emissions—essentially minimising emissions that are directly caused by their own operations and the energy consumed during those operations. But to accelerate their carbon reduction, chip companies must also target emissions from upstream Scope 3 activities. These primarily encompass CO_2 output contributed by suppliers; in other words, raw materials, manufacturing, and the like related to supplier production.

Upstream Scope 3 comprises about 40% of chip makers' carbon emissions (excluding Scope 3 downstream emissions, covering the actual use of a chip by customers).

According to McKinsey's report, the majority—80% of GHG emissions from semiconductor manufacturing— is either directly from the fabs during production (Scope 1) or a byproduct of the generation of electricity, steam, heating, and cooling equipment (Scope 2). Note that the Scope 1 emissions are a consequence of using process gases that have a high global warming potential (GWP) like Perfluorocarbons (PFCs), Hydrofluorocarbons (HFCs), $NF_{3,}$ and N_2O.

Thus, while semiconductor chip makers have taken steps to reduce Scope 1 and Scope 2 emissions to accelerate their carbon-reduction efforts, chip companies are also going well beyond that, targeting Scope 3 emissions from upstream suppliers and driving a holistic transformation that includes:

- Making long-term purchasing agreements for decarbonised products
- Reconfiguring the supply chain to focus on sustainability
- Collaborating with suppliers to develop emissions-reduction strategies

Corporate initiatives in the semiconductor industry to address climate change are focused on achieving net-zero emissions by 2050, driven by intense energy consumption in chip fabrication and rising demand for advanced computing. Key strategies include transitioning to renewable energy, implementing advanced gas abatement technologies, reducing water consumption, and enhancing supply chain sustainability.

9.1.6 The Indian Semiconductor Landscape

The semiconductor industry, globally acknowledged as a critical pillar of technological advancement, is witnessing a transitional phase in India.

We consider the current state of India's semiconductor landscape, highlighting the industry's evolution, the pivotal role of government initiatives, and the initiatives by key market players.

9.1.6.1 Improving product quality and longevity for sustainable development

India is integrating sustainability into its semiconductor industry by promoting green manufacturing, water conservation, and renewable energy usage through the India Semiconductor Mission (ISM) and the Semicon India Programme. With a ₹ 76,000 crore outlay, the initiative focuses on creating high-tech clusters with infrastructure for reducing carbon footprints and managing high energy/water consumption.

9.1.6.2 Initiatives for India's Sustainable Semiconductor Production

1. **Integration of Renewable Energy and Energy Efficiency:** Where practical, Indian factories are switching to renewable energy sources and implementing energy-efficient technology. Thus, solar panels, microgrids, and energy-efficient lighting are being integrated into factories. The shift to sustainable energy sources is also being aided by cooperative initiatives with technology and electricity suppliers.
2. **Water Conservation and Recycling:** Effective water management has become a top concern due to the high water requirements for semiconductor fabs. Up to 75% of the water used by advanced factories is treated and reused through recycling technologies. In water, such steps are necessary to continue operations and prevent the depletion of local water supplies in limited locations, especially in states like Karnataka and Maharashtra, where semiconductor production is growing.
4. **Cutting Carbon Emissions with Cutting-Edge technology:** Factories are investing in process-gas abatement technologies and less environmentally harmful substitute materials as they strive for fewer emissions. India's Fabs are progressively putting modern air and chemical filtration systems into place as part of their commitment to lowering greenhouse gas emissions.
5. **Worker and Supply Chain Development:** India is making great strides in developing a semiconductor ecosystem that prioritises domestic sourcing, lessens reliance on imports, and enhances worker competencies. To promote the local manufacturing of semiconductor components, the government has launched a number of programmes, such as the Production-Linked Incentive (PLI) scheme. This programme is a component of India's larger effort to build a more robust supply chain that reduces transportation-related emissions while simultaneously addressing interruptions.

Thus, initiatives comprise minimising emissions in Scope 1 and Scope 2 as well as Scope 3 emissions.

In parallel, industry, academia, and governmental organisations are working together to assist initiatives to develop a competent workforce. The Electronics Sector Skills Council of India (ESSCI) and other organisations are essential to training employees in semiconductor manufacturing technology in order to address the skills gap. In the upcoming years, this will guarantee that India has the people resources needed to fulfil the rising demand for semiconductors.

9.1.6.3 Circular Economy Initiatives and E-Waste Management:

E-waste significantly contributes to greenhouse gas (GHG) emissions—estimated to be 1.4%-5.9% of global emissions—through energy-intensive manufacturing, inefficient recycling, and improper disposal. Burning and landfilling electronics release carbon dioxide, methane, and hazardous refrigerants (HFCs).

In keeping with sustainable growth, India is also concentrating on e-waste management in the semiconductor sector and Circular Economy practices.

The government has started measures to promote the recycling of semiconductor components and other electronic debris, like the National E-debris Management Policy and other state-level e-waste collecting programmes. By encouraging the appropriate disposal and recycling of semiconductor materials, these regulations seek to lessen the environmental impact of the industry.

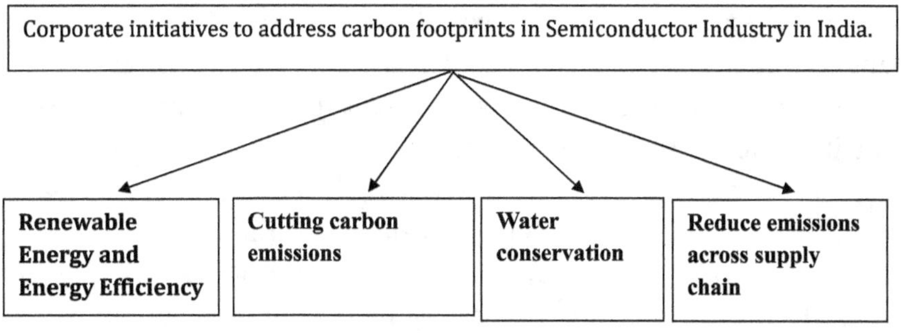

Figure 9.1 Corporate initiatives to address carbon footprints in the Semiconductor Industry in India.

9.2 Case 2: Corporate initiatives to reduce the carbon footprint of the textile sector

9.2.1 The textile industry is responsible for a significant amount of global CO_2 emissions.

In fact, textiles and clothing constitute one of the most polluting industries, caused by high energy, water and chemical use, the generation of textile waste and microfibre shedding into the textiles and clothing during laundering (Niinimäki et al., 2020). Due to lengthy supply chains and energy-intensive production methods, apparel and footwear industries generate 8–10% of global carbon emissions. Furthermore, it has been estimated that up to 20% of industrial wastewater pollution is caused by textile dyeing and finishing (Morlet et al., 2017).

Textile production occurs through geographically long and complex supply chains that include growers and processors of raw fibres, yarns and textiles, weavers, knitters, dyers and finishers, product manufacturers and distributors. While textiles have various end uses, the sector is essentially fashion-driven, as most global fibre production (60%) is destined for clothing (Niinimäki et al., 2020). The production and manufacturing of natural and synthetic textiles continue to be the most ecologically damaging aspects of the industry.

Major textile production countries such as China, India and Bangladesh still rely on coal. Logistical requirements of transporting goods globally at the various stages of production and distribution contribute to textiles' carbon footprint, alongside energy deployed in garment care such as washing, drying, pressing and dry-cleaning. Finally, the deterioration of textile waste in landfill and/or incineration causes hazardous chemicals and greenhouse gases (GHG) to be released into the environment (Juanga-Labayen et al., 2022). Textiles' ecological footprint has been exacerbated by the growth of fast fashion - cheaply manufactured, low-quality clothing designed to be trendy and purchased frequently - which results in a waste of resources and greater textile waste as the clothing is more quickly discarded. Clothing production has doubled since 2000 and shows no sign of abatement due to globalisation, urbanisation and population growth, which fuel demand for more clothing (Filho et al, 2022).

9.2.2 The Textile Industry and Climate Change

Most textile and clothing production takes place in Global South countries which rely heavily on fossil fuels for electricity supply, thereby increasing the carbon footprint of each clothing item produced. Because garments are worn for a shorter amount of time before they end up in landfill or are incinerated, carbon emissions have increased from both manufacturing and the production of materials used by the fashion industry.

The use of synthetic fibres has grown exponentially, and polyester has now surpassed cotton as the most widely used fashion fabric (Niinimäki et al., 2020). However, carbon emissions from synthetics are much higher than those of cotton because they are made from fossil fuels such as unrefined petroleum products, thereby contributing to climate change through carbon emissions.

The drive to produce more fabrics to meet the rising manufacturing demand of the fashion industry also contributes to deforestation. Rayon (a semi-synthetic fibre made from purified cellulose) and similar fabrics such as viscose and modal are manufactured from wood pulp. The rayon/viscose industry is projected to expand in response to increasing demand (Mordor Intelligence, 2021), which will likely amplify the threat to forests globally. An estimated 30% of rayon and viscose utilised by the fashion industry are harvested from ancient and endangered forests (Mordor 2021). Deforestation, the second-largest driver of GHG emissions globally, has accelerated the severity of extreme heat in temperate regions of North America and Europe (FAO, 2018).

As a result of fast fashion and shortening product life cycles, post-consumer textile waste contributes significantly to the increase of landfills and incineration (causing significant emissions of toxic gases). Without a strategy of recovery and recycling, the consequences for the environment will worsen, bearing in mind the quantity of resources, water and energy that go into the production of new textile products, which mainly become textile waste once disposed of by the consumer.

9.2.3 Corporate Initiatives to address climate change in the textile industry

To address climate change in the textile industry, corporations are implementing a range of initiatives, including shifting to renewable energy, using sustainable materials, reducing water consumption, streamlining logistics, minimising waste, and promoting circular business models.

These efforts aim to reduce the industry's carbon footprint and transition towards more sustainable practices:

1. Shifting to Renewable Energy:
   - Investing in Renewable Energy Systems:

     Textile manufacturers are investing in solar panels, wind turbines, and other renewable energy systems to reduce their reliance on fossil fuels and lower carbon emissions.
   - Energy Efficiency:

     Implementing energy-efficient technologies and practices can also significantly reduce energy consumption and emissions.
2. Utilising Sustainable Materials:
   - Natural Fibres:

     Promoting the use of natural fibres like organic cotton, wool, and hemp, which have lower environmental impacts than synthetic fibres. Synthetic fibres are a significant source of greenhouse gas (GHG) emissions. They are derived from petrochemicals (oil and natural gas) and their production, use, and disposal release large amounts of carbon dioxide and other pollutants.
   - Sustainable Manufacturing:

     Focusing on sustainable manufacturing processes that minimise resource use, renewable energy and also minimise waste.

3. Reducing Water Usage:
   - Water-Efficient Technologies:

     Implementing water-efficient technologies and processes to reduce water consumption in the textile production process.
   - Water Reuse and Recycling:

     Exploring options for reusing and recycling water to further minimise water usage.
4. Streamlining Logistics:
   - Nearshoring/Reshoring:

     Shortening supply chains through nearshoring or reshoring production to reduce **transportation emissions**. Nearshoring moves production to nearby countries which may have lower labour costs with faster, simpler logistics, while reshoring brings manufacturing back to the home country for maximum control and quality assurance. Both strategies, often driven by the need for supply chain resilience, shorter lead times, and reduced geopolitical risk, offer faster, more agile alternatives to traditional, distant offshoring.

- Digital Product Development:

  Utilising digital design tools and visualisation technologies to reduce the need for physical production and transportation.

5. Reducing Waste:

- Waste Reduction Strategies:

  Implementing strategies to minimise waste throughout the production process, including using less material and reducing scrap. Textile waste is a major contributor to greenhouse gas (GHG) emissions, with the fashion industry responsible for approximately **8–10% of global carbon emissions.**
- Recycling and Circularity:

  Promoting recycling and circular business models, including the use of recycled materials and the development of innovative end-of-life solutions.

6. Promoting Circular Business Models:

- Design for Durability and Repairability:

  Designing products that are durable, repairable, and reusable to extend their lifespan and reduce waste.

- Circular Economy Initiatives:

  Supporting and participating in a circular economy involves initiatives that promote the reuse and recycling of textile materials. In a circular economy, products and materials are kept in circulation through processes like maintenance, reuse, refurbishment, remanufacture, recycling, and composting. The circular economy tackles climate change and other global challenges, like biodiversity loss, waste, and pollution.

7. Other Initiatives:

- Chemical Footprint Reduction:

  Exploring measures to reduce the use of chemicals in textile production and adopt eco-friendly alternatives.

- Sustainability Certifications:

  Encouraging the use of sustainability certifications to ensure transparency and accountability in textile production.

- Collaboration and Partnerships:

  Engaging in collaborations and partnerships with other companies, organisations, and governments to promote sustainable practices across the industry.

9.2.4 Textile Industry and Climate Change in India

The Indian textile industry is a major global player, contributing about 2 to 3% of India's GDP. This industry is highly vulnerable to climate change, particularly through reduced cotton yields caused by heatwaves and erratic rainfall. Conversely, the sector is a major polluter, relying on fossil fuels for thermal energy and causing significant water pollution. (Global Efficiency Intelligence)

Impact of Climate Change on India's Textile Industry

- Raw Material Disruptions: Climate-induced stressors like heatwaves and droughts are projected to reduce cotton yields by 268 to 277 kgs per hectare by the 21st century. In 2022, extreme weather caused a ~80% drop in cotton arrivals in Punjab, heavily impacting ginning units. Ginning is the process of separating cotton fibres (lint) from seeds and removing debris like leaves and stems.
- Livelihood Vulnerability: Erratic rainfall and floods directly threaten the livelihoods of millions of smallholder cotton farmers.
- Operational Risks: Intense heatwaves, such as those in Surat, create dangerous working conditions and operational challenges for workers in steam-based, high-temperature manufacturing processes.

Textile Industry's Contribution to Climate Change in India

- High Carbon Footprint: The sector relies heavily on coal and petroleum, which account for ~46% of energy use, primarily for thermal energy in dyeing, drying, and finishing processes.
- Water and Chemical Pollution: The industry is a major source of wastewater, accounting for ~20% of industrial water pollution due to toxic chemical dyes.
- Waste Generation: Only 1% of fibres are recycled, with fast fashion trends contributing to significant textile waste and pollution.

Transition to Sustainability

- Electrification: Electrifying thermal processes, which account for 43% of energy consumption, could lead to 25–30% efficiency gains and significantly reduce emissions. But the process should use **renewable energy** as much as possible.
- Sustainable Practices: The sector is shifting towards renewable energy, water recycling, and adopting eco-friendly technologies, supported by government initiatives like modernisation schemes.

- Support for Small Players: There is a need to support MSMEs through technology upgrades and capacity building to manage the costs of adopting green, sustainable practices.

9.2.5 Corporate Initiatives to Address Climate Change in India's Textile Industry

These initiatives focus on reducing carbon footprints through renewable energy adoption, water recycling, and sustainable sourcing. Key actions include installing solar power, implementing Zero Liquid Discharge (ZLD) systems, using recycled materials (rPET) for packaging, and adhering to ESG standards to meet global, especially EU, regulations.

Key Corporate Initiatives:

- Renewable Energy & Efficiency: Companies like Welspun Group and others are installing solar rooftop systems and shifting to biomass-based energy to replace fossil fuels.
- Water Management & ZLD: To combat high water consumption, textile hubs, particularly in Bengaluru, have implemented Zero Liquid Discharge technology, treating and recycling wastewater.
- Sustainable Materials: Major players are increasing the use of organic cotton, recycled polyester (rPET) from plastic bottles, and eco-friendly dyes to reduce environmental and GHG impact.
- Circular Economy Practices: Adoption of 'Project SU.RE' by 30 major Indian fashion firms commits to using sustainable raw materials by 2025. Companies are also focusing on recycling post-consumer garments.
- Digital ESG Tracking: Firms are employing digital platforms to measure and report on environmental, social, and governance (ESG) KPIs, ensuring compliance with international trade demands.
- Waste Management: Initiatives include using refuse-derived fuel (RDF) from textile waste to power boilers, reducing landfill waste. **Refuse-Derived Fuel (RDF) produces greenhouse gases (GHGs) when burned, but it is generally considered a lower-carbon alternative to coal and a tool to reduce overall emissions compared to landfilling.**

- Streamlining logistics by adopting digital technologies, automating warehouses, and improving supply chain visibility to handle fast fashion cycles and reduce emissions as well as costs.

Leading firms such as Welspun Living, Arvind Mills, Shahi Exports, and Sutlej Textiles are at the forefront of these efforts to align with India's climate goals and international standards.

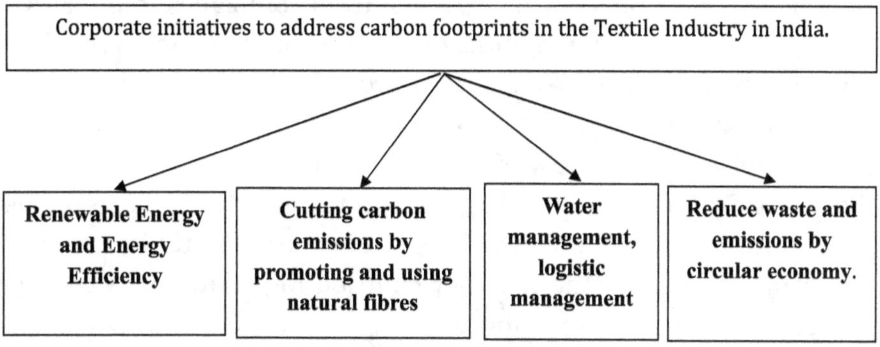

Figure 9.2 Corporate initiatives to address carbon footprints in the Textile Industry in India.

9.3 Case 3: FMCG industry and climate change

The FMCG industry is characterised by high-volume sales, quick inventory turnover, and various products catering to consumer needs. These goods include essential everyday items such as food and beverages, toiletries, cleaning supplies, and other low-cost household items.

9.3.1 Types of FMCG

FMCG products are typically divided into various types depending on the sector in which they are sold. These consist of the following:

- Food and Beverages
- Personal Care
- HealthCare Goods
- Home Care Commodities

The fast-moving consumer goods (FMCG) sector and climate change are deeply intertwined – the sector is a major cause of GHG emissions, but is also heavily impacted by it. Much of global GHG emissions comes from food production, with food waste and deforestation also being major contributors to GHG emissions.

Greenhouse gas (GHG) emissions, which are one of the major causes of rising global warming, come from companies in the FMCG sector, accounting for a significant amount of the global emissions. This industry produces over one-third of global greenhouse gas (GHG) emissions, with 80%–95% stemming from Scope 3 supply chain activities, such as raw material sourcing and agriculture. Carbon dioxide, CO_2, Methane, & NOx gases form a major part of greenhouse gases released. Due to rising GHG emissions from FMCG companies, it became important among major contributing sectors to report gas emissions during their operations.

9.3.2 Carbon Footprints Across the FMCG Supply Chain

Carbon footprints in FMCG are the total set of greenhouse gas emissions caused at various steps in the supply chain. The decarbonising process in a FMCG supply chain encompasses best practices in reducing carbon emissions across the supply chain, from materials sourcing through product design, manufacturing, distribution, delivery and finally, end-of-life recycling.

Typical supply chain management requires organisations to optimise various metrics ranging from costs, levels of service, inventory levels and more. Organisations that pursue a green model of business must be mindful of CO_2 emissions in their operations. At each step, there is carbon generation. Figuring out ways to make plants and distribution sites more energy-efficient and to decrease distance in the transportation network can help to achieve two goals - a leaner, more cost-effective supply chain and a reduced carbon footprint.

The integrated supply chain, which will achieve a reduction in carbon footprint, will emphasise renewable fuel sources, as non-renewable sources become more scarce, polluting and expensive. There are many different technologies being developed for different applications, including those based on hydrogen, solar, wind, and ethanol. However, renewable sources of energy often have some significant barriers to being used widely in the supply chain as well as in industry in general. Using solar energy is still relatively expensive (though prices are expected to fall) and solar cells still require fossil fuels. Hydrogen power is challenging to produce efficiently on a large scale, difficult to store, and difficult to transport.

While wind power is clean, its cost remains relatively high for the resultant (and often intermittent) power output. If and when these and/or other alternative fuels become scalable and cost-competitive, they will play critical roles in powering future supply chains (Beamon, 2008).

Companies from the FMCG sector are under increasing pressure to reduce their carbon footprint. Consumers are becoming more aware of excess carbon dioxide emissions' detrimental environmental effects. Their consumption behaviour changes due to this, and they take an active interest in sustainability issues.

As referred to earlier, greenhouse gas (GHG) emissions are generally classified into Scope 1, 2, & 3 emissions. Scope 1 includes direct GHG emissions from stationary fuel combustion and refrigerant gases. Thus, Scope 1 emissions are "direct emissions" from sources that are owned or controlled by the company. Scope 2 encompasses indirect GHG emissions from the on-site generation and purchase of electricity. These are called indirect emissions. Scope 3 involves indirect GHG emissions in the company's entire value chain, such as the purchase of goods and services, upstream transport and distribution, and employee commute.

(https://www.globaldata.com/data-insights/consumer/total-ghg-emissions-by-major-fmcg-companies-2096157/)

The journey towards sustainability within the FMCG realm commences with a fundamental re-evaluation of product design and packaging strategies. Today, companies are embarking on a transformative path by using eco-friendly materials, including biodegradable plastics, recycled paper, and reusable containers.

Limiting the carbon footprint of a fast-moving consumer goods (FMCG) company is crucial to addressing climate change and promoting sustainability. Here are several strategies an FMCG company can adopt to reduce its carbon footprint:

9.3.3 Strategies to reduce carbon footprint in the FMCG Industry

- Sustainable sourcing: The FMCG companies must ensure the procurement of raw materials from sustainable and environmentally responsible sources. They should also choose suppliers that prioritise renewable resources, ethical practices, and minimise transportation emissions.
- Energy efficiency: The FMCG companies must implement energy-efficient practices within manufacturing facilities, such as using LED lighting, optimising heating, ventilation, and air conditioning (HVAC) systems, and investing in energy-efficient machinery. They should also consider renewable energy sources, such as solar or wind power, for operations.
- Packaging optimisation: The FMCG companies must minimise packaging waste by using lighter and more eco-friendly materials, such as recycled or biodegradable packaging. Explore innovative packaging designs that reduce the overall amount of packaging required and encourage recycling.
- Supply chain optimisation: The FMCG companies must optimise logistics and transportation by streamlining routes, implementing efficient loading and unloading procedures, and considering alternative transportation methods, such as rail or waterways. They must collaborate with logistics partners to reduce the carbon footprint across the entire supply chain.
- Waste management: Implement robust waste management systems that promote recycling and composting. Minimise waste generation and encourage employees and customers to recycle through proper education and awareness campaigns.
- Product lifecycle assessment: The FMCG companies must conduct a thorough assessment of the entire product lifecycle, including raw material extraction, production, distribution, use, and disposal...

Identify opportunities to reduce emissions at each stage of the product's lifecycle, and implement measures accordingly.

- Carbon offsetting: The FMCG companies must invest in carbon offset projects to compensate for unavoidable emissions. Support initiatives such as reforestation, renewable energy projects, or methane capture projects that help offset the company's carbon emissions.
- Employee engagement and education: The FMCG companies must foster a culture of addressing climate change and sustainability among employees by providing training and education on environmental issues, inviting ideas for improvement, and involving them in sustainability initiatives. They must encourage telecommuting or flexible work arrangements to reduce commuting-related emissions.
- Collaborate with stakeholders: The FMCG companies must engage with customers, suppliers, and other stakeholders to promote sustainable practices and encourage them to reduce their carbon footprints. They must collaborate with industry peers to share best practices and drive collective efforts to tackle climate change.
- Continuous improvement and measurement: The FMCG companies must regularly monitor and measure their carbon footprint to identify areas for improvement. They must set specific, measurable, achievable, relevant, and time-bound (SMART) targets, and regularly review progress. They must also adopt new technologies and innovations that help further reduce the carbon footprint.

Implementing these strategies will help the FMCG company make significant progress towards limiting its carbon footprint, thereby contributing to a more sustainable future while enhancing its brand reputation as an environmentally responsible organisation.

(https://www.globalpraxis.com/insights/fmcg-managing-your-carbon-footprint)

Given the current state of climate change, it is now critical for the FMCG industry to take concrete steps towards **carbon neutrality**. By reducing and offsetting carbon emissions and prioritising sustainable practices throughout their value chain, FMCG brands can make a significant environmental difference and have a substantial impact in the fight for net-zero.

By embracing carbon neutrality, FMCG companies can not only fulfil their environmental responsibilities but also position themselves as leaders in sustainability, shaping a more sustainable future for all.

9.3.4 Carbon Neutrality

Carbon neutrality means having a balance between the amount of carbon emitted and what is absorbed from the atmosphere. Companies and their products become carbon neutral when they can accurately report on the carbon emissions they have produced and reduce this via carbon offsetting projects.

The FMCG industry contributes a significant amount of carbon emissions because we use these products as part of our everyday lives. The World Economic Forum states that the food and FMCG industries together produce more than one-third of global emissions, which mainly come from their upstream supply chains.

This is a complex industry to track emissions and environmental impact for because each product will vary; for instance, textile production contributes more than 1 billion tonnes of CO_2e per year. On top of this, each region will also produce different amounts of CO_2 and have its own regulations and goals, which one can see in the data available in the industry reports. Nonetheless, it is important for all areas of the FMCG industry to get on board with net-zero goals and make a positive change.

(Source: https://ourworldindata.org/co2-and-greenhouse-gas-emissions)

9.3.5 How Can FMCG Companies Achieve Carbon Neutrality?

Carbon neutrality is a critical goal for FMCG companies to try to reduce their carbon footprint on the planet. Every brand and supplier within the value chain should be taking proactive measures to reduce their energy consumption, choose more environmentally friendly materials which do not produce greenhouse gases, re-evaluate their current processes and implement sustainability strategies such as investing in renewable energy. All of this will help companies work towards carbon neutrality.

(Source: https://ourworldindata.org/co2-and-greenhouse-gas-emissions)

By adopting carbon neutrality, FMCG companies not only fulfil their environmental responsibilities but also position themselves as leaders in sustainability, shaping a more sustainable future for all.

(https://marketplace.adec-innovations.com/blogs/why-achieving-carbon-neutrality-is-so-critical-for-fmcg/)

9.3.6 FMCG Companies and Climate Change in India

India is currently the third-largest emitter of greenhouse gases globally. Dependence on fossil fuel imports has impacted the country's energy security and economic growth. Forest degradation is leading to increasing greenhouse gas emissions and loss of livelihoods.

9.3.6.1 Government of India Climate Priorities

India has demonstrated a sustained commitment to combating climate change, and is one of the few parties to the Paris Agreement, requiring countries

(1) to commit to reduce greenhouse gases,

(2) to have adopted emissions-reduction targets that are compatible with a two- degree global warming scenario.

Top goals under India's Nationally Determined Contributions to the Paris Agreement include:

- Expand the power sector's reliance on non-fossil fuels to 40 percent by 2030;
- Reduce emissions intensity of GDP by one-third over the 2005 baseline;
- Create an additional carbon sink of 2.5 to 3 billion tons of CO2 through additional forest and tree cover by 2030;

- Invest in vulnerable sectors such as agriculture and water in key regions; and
- Address resource gaps by mobilising domestic and international resources to implement climate mitigation and adaptation actions.

(https://www.usaid.gov/climate/country-profiles/india)

The commitment to reduce carbon emissions **also translates hugely to the FMCG industry**.

In **general, reducing carbon emissions in India's Fast-Moving Consumer Goods (FMCG) industry involves a strategic shift towards renewable energy, sustainable packaging, and greener supply chain logistics.**

Key initiatives include utilising solar/wind energy, adopting Post-Consumer Recycled (PCR) materials, optimising cold chains, and implementing energy-efficient manufacturing to meet sustainability goals.

9.3.6.2 Some strategies an FMCG company can adopt to reduce its carbon footprint:

- Sustainable sourcing: Ensuring the procurement of raw materials from sustainable and environmentally responsible sources. Choosing suppliers that prioritise renewable resources, ethical practices, and minimise transportation emissions.
- Energy efficiency: Implementing energy-efficient practices within manufacturing facilities, such as using LED lighting, optimising heating, ventilation, and air conditioning (HVAC) systems, and investing in energy-efficient machinery. Consider renewable energy sources, such as solar or wind power, for operations.

(HVAC stands for Heating, Ventilation, and Air Conditioning. It refers to the comprehensive system of components designed to control indoor temperatures, humidity, and air quality in residential and commercial buildings. Systems commonly feature furnaces, heat pumps, air conditioners, and air handlers to create a comfortable, healthy environment)

- Packaging optimisation: Minimising packaging waste by using lighter and more eco-friendly materials, such as recycled or biodegradable packaging. Exploring innovative packaging designs that reduce the overall amount of packaging required and encourage recycling.
- Supply chain optimisation: Optimising logistics and transportation by streamlining routes, implementing efficient loading and unloading procedures, and considering alternative transportation methods, such as rail or waterways. Collaborating with logistics partners to reduce the carbon footprint across the entire supply chain.
- Waste management: Implementing robust waste management systems that promote recycling and composting. Minimising waste generation and encouraging employees and customers to recycle through proper education and awareness campaigns.
- Product lifecycle assessment: Conducting a thorough assessment of the entire product lifecycle, including raw material extraction, production, distribution, use, and disposal. Identify opportunities to reduce emissions at each stage of the product's lifecycle.

9.3.6.3 Some green supply chain initiatives taken by FMCG companies in India

(Mazumder and Chatterjee, 2007, Verma, Malhotra and Rao, 2017):

On the warehousing front, the warehouse contractors are advised to work on better insulation of the walls of the warehouse so that it leads to less energy consumption.

Replacing the lighting systems with energy-efficient lights. LEDs, smart and movement-sensing lighting systems, can lead to energy savings of 90%.

Since transportation is a major cost component in the overall logistics cost of a FMCG company, organisations are working on improving the fuel efficiency of the trucks. The companies are also trying to reduce the idling time of trucks in transit. They are working on pickup strategies for trucks while delivering.

A paper and packaging conglomerate has taken a step further, wherein they have been engaging local communities to grow trees according to the demands of the local ecosystem, which they later consume in their paper and packaging plants.

A major food company begins using spent coffee grounds and biogas from wastewater to fuel its boilers and enlists its entire supply chain to reduce energy consumption and cost within a global logistics initiative.

Thus, when the company optimises the whole supply chain and not some of its parts, the value delivery of the supply chain will be greatly enhanced in terms of tangible and intangible social and environmental benefits and also address major issues such as climate change. Awareness among the people comes in, and CSR (corporate social responsibility) comes into light, which involves every responsibility an organisation should have for society, community, environment and customers.

9.3.7 Strategies to reduce emissions across the supply chain in Indian FMCG companies

9.3.7.1 Inbound Logistics

Indian FMCG companies have taken initiatives to Reduce Carbon Footprints in **Inbound Logistics Procurements from suppliers. Sourcing plays a major role in reducing carbon** footprints in the supply chain. Purchasing from suppliers who are energy-efficient, who use organic fertilisers, and use recyclable materials and materials that are biodegradable and use more energy-efficient processes and less toxic materials will reduce carbon footprints. This initiative encompasses reducing carbon footprint contribution of varied raw materials used, investing in transportation of raw materials from suppliers to the company to reduce carbon emissions, working in collaboration with suppliers to reduce carbon emissions, etc.

9.3.7.2 Initiatives to Reduce Carbon Footprints in Outbound Logistics

This initiative encompasses investing in the transportation of products from the company to retailers/ consumers to reduce carbon emissions, measuring carbon emissions from various modes of transportation, planning out the usage of each vehicle to its best in back-and-forth trips in terms of capacity and technology used, with the intention of reducing carbon footprints.

9.3.7.3 Initiatives to Reduce Carbon Footprints in Production or Internal Logistics

There are many ways to reduce carbon in a manufacturing plant. A closer look at the power source is an excellent start. For example, two plants that consume the same amount of energy may have completely different carbon footprints. That's because one uses hydro-generated electricity while the other operates with coal-generated electricity. Looking beyond the source of energy, there are ample opportunities to reduce consumption. These could be evaluating how water is consumed as a resource throughout the production stages, how efficient the plant layout and production plans are, how much inventory is produced in excess of what is needed. Streamlining production steps and reducing toxic materials and harmful emissions can each have a significant impact on how green the supply chain is. This is done by implementing Zero-Emission Strategy and complying with Six Sigma methodology, etc.

9.3.7.4 Initiatives to Reduce Carbon Footprints in Storage/ Distribution Centres

The operations involved in the storage of FMCG goods are an integral step in the supply chain and contribute to the carbon footprint calculation and greenhouse gas emissions. For example, if the storage is used for chilling the goods, we can calculate the emissions caused over that period of time and attribute the corresponding amount of emissions to those goods. Before leaving the distribution centre, the goods get disaggregated into cases and then shipped to different retailers. The cases may have differing footprints from this point because of the different transportation distances and the different cooling times. A distribution centre generates different and extra distances for the delivery of goods to different destinations, which in turn leads to extra costs and high carbon emissions in transportation.

It is proposed that the location of additional storage capacities should be in the vicinity of existing production facilities, and that production capacities should be concentrated. This reduces the carbon emissions to a great extent.

9.3.7.5 Initiatives to Reduce Carbon Footprint in Packaging

Large packages take more energy to produce. The larger the package, the less shipment consolidation is possible. Fortunately, new packaging materials and designs make smaller package volumes possible while providing the same level of strength and product protection. Improved package designs, such as using biodegradable substances to package, will also reduce overall carbon footprints. "(Exploring the Green Supply Chain management: A Technical Review, 2013)".

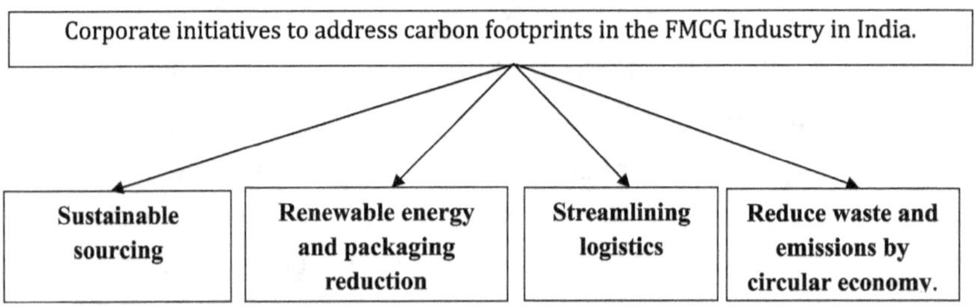

Figure 9.3 Corporate initiatives to address carbon footprints in the FMCG Industry in India.

9.4 Case 4: Electric vehicles industry and climate change impacts

In a world increasingly focused on reducing carbon dioxide (CO2) emissions and other kinds of pollution, electric vehicles (EVs) are becoming more and more attractive. Research consistently shows EVs produce fewer greenhouse gases and air pollutants over their lifetime compared to petrol and diesel cars—even when factoring in vehicle production and electricity generation.

In order to address climate change, various measures are being incorporated across all continents, including reducing the generation of greenhouse gases. Since electric vehicles, EVs, run on electricity and generate much less GHGs, EVs are becoming a preferred choice with consumers all across the world.

Why people are moving to use electric vehicles, EVs, instead of regular vehicles is because electric vehicles run on electricity and generate much less greenhouse gases. However, it has been a relatively new concept, and though the sale of electric vehicles has been consistently going up, there are still some hesitations in the minds of consumers about buying electric vehicles. These hesitations pertain to the availability of charging stations, driving range, maximum speed, initial cost, maintenance cost, etc.

9.4.1 Benefits of Using Electric Vehicles

Reduced greenhouse gas emissions

One of the greatest benefits of EVs is their ability to drastically reduce greenhouse gas emissions. Unlike petrol and diesel cars, EVs produce zero tailpipe emissions, meaning they don't release any CO2 into the atmosphere. This massively improves air quality, creating cleaner streets for all. (Lawrence 2025)

In the case of traditional petrol cars, these emit around 165 grams of CO2 per kilometre, and diesel cars emit about 170 grams of CO2 per kilometre. In contrast, EVs average just 50 grams of CO2 per kilometre, including consideration of the electricity generated. Therefore, over one year, just one EV can save an average of 1.5 million grams of CO2. This reduction in CO2 emissions highlights the massive environmental advantage of EVs over petrol and diesel cars.

Reduced noise pollution

Unlike petrol and diesel cars, where noise is generated from the engine and exhaust, EV motors are almost silent, especially at lower speeds. This results in quieter streets, which can make urban areas more peaceful and pleasant for both people and wildlife.

Improved energy efficiency

EVs are more energy-efficient than petrol and diesel cars through converting a higher percentage of energy from the car battery into motion. While petrol cars waste approximately 75–84% of their energy through heat, friction, and other inefficiencies, EVs only waste about 31-35%. Therefore, EVs help reduce the overall energy demand by losing less energy. This makes EVs a smarter, greener choice for the environment.

9.4.2 EV Production Emissions Contribute to Climate Change

EVs are more energy-efficient than petrol and diesel cars through converting a higher percentage of energy from the car battery into motion. While petrol cars waste approximately 75–84% of their energy through heat, friction, and other inefficiencies, EVs only waste about 31-35%. Therefore, EVs help reduce the overall energy demand by losing less energy. This makes EVs a smarter, greener choice for the environment.

However, there are challenges to using EVs too.

Though EVs have tremendous benefits in reducing emissions, producing EVs does use a lot of energy. The emissions created during EV production tend to be higher than those of petrol cars. This is due to the manufacturing of lithium-ion batteries, an essential part of an EV, which releases several greenhouse gas emissions. These include CO2, methane (CH4), sulphur dioxide (SO2), and nitrogen oxides (NOX).

However, as technology advances, this is changing for the better. With more efficient manufacturing techniques – such as the use of sustainable and recycled materials – being introduced, the volume of greenhouse emissions created during the production of batteries will improve.

9.4.3 Battery Recycling Dilemmas

The waste and recycling of EV batteries present significant environmental challenges. EV batteries can last up to 20 years, but when they eventually need replacing, it can result in substantial waste. Recycling these batteries is particularly difficult due to their complex design, high costs, and low recovery rates of harmful materials like lithium and cobalt. Also, improper handling or disposal can exacerbate the problem. For example, hazardous substances in EV batteries can leach into soil and water, posing serious risks to ecosystems and public health.

However, reusing and recycling EV batteries is an ever-growing market. Research into the use of second-hand batteries is looking at ways to reuse batteries in new technologies such as electricity storage. This could help balance the National Grid and store renewable energy, reducing the environmental impact of EV battery production. Also, due to increased energy storage, renewable energy can become more reliable - even when the sun isn't shining or the wind isn't blowing, there will still be a steady supply of renewable energy available!

Finally, with advancing vehicle-to-home (V2H) technologies that power homes, one day, we could all have our own energy storage units. This not only provides backup power and reduces energy costs but also creates another opportunity to reuse EV batteries. Repurposing batteries for home energy storage extends their lifecycle, minimises waste, and lessens the environmental burden of manufacturing. This circular approach supports a more sustainable battery lifecycle, counteracting current environmental damage caused by production.

Overall, even after considering lithium-ion battery manufacturing, EVs are still a greener option. Their lifetime greenhouse gas emissions are significantly lower than those of petrol and diesel vehicles, making them a critical tool in reducing global carbon emissions.

9.4.4 Mining Impacts:

When producing EV batteries, the mining of required materials such as lithium, cobalt, and nickel causes multiple environmental issues. For example, refining nickel and cobalt can increase sulphur dioxide (SO2) emissions, which creates pollution hotspots. Similarly, toxic chemicals used in mining, such as hydrochloric acid, can contaminate water supplies and soil. In some cases, this has led to incidents like fish die-offs in rivers near mining sites.

Also, mining for EV battery materials can have severe social consequences, particularly in regions where cobalt mining needs to be done and cobalt mining has been linked to poor working conditions, where miners often face hazardous environments and inadequate safety measures. Additionally, there are significant concerns about child labour and the violation of indigenous communities' rights, highlighting the urgent need for ethical and sustainable EV battery mining practices to be introduced.

However, research is making significant strides in developing EV battery technologies that use fewer harmful materials, paving the way for more sustainable solutions. Simultaneously, advancements in EV battery recycling now allow up to 90% of battery materials to be recovered and reused. This minimises waste and conserves resources.

Companies are also taking steps to prioritise ethical sourcing, ensuring fair labour practices and eliminating child labour. For example, the Global Battery Alliance (GBA), which includes UK-based companies and organisations, is committed to ending child labour in the battery supply chain through various measures, such as developing a digital Battery Passport to track and authenticate the absence of child labour in the sourcing of cobalt. Meanwhile, international cooperation is driving efforts to adopt sustainable mining practices and secure a reliable supply of critical materials.

While the production of EV batteries continues to pose significant environmental and social challenges, ongoing advancements in technology, recycling, and ethical sourcing are critical to mitigating these impacts. By confronting these issues and fostering sustainable practices, the industry can ensure that the shift towards EVs leads to a cleaner future without compromising human rights or the health of our planet.

In order to address climate change, various measures are being incorporated across all continents, including reducing the generation of greenhouse gases. Since electric vehicles run on electricity and generate much less GHGs, EVs are becoming a preferred choice with consumers all across the world.

9.4.5 Electric Vehicles in India

Electric vehicles have become very popular all over the world and now in India as well. These vehicles are run by electricity and not by gasoline. These vehicles use electric motors running on rechargeable batteries. This gives rise to much reduced emissions of greenhouse gases and thus addresses the climate change problem.

In addition to the tremendous benefits offered by electric vehicles in reducing emissions, thereby addressing climate change, **electric vehicles have much lower running costs** because the batteries can be charged at home. Even if they are charged at public charging stations, the associated cost is quite low.

The driving range, distance it can run after a charge, offered by electric vehicles is also comparable to most gasoline-driven vehicles.

India's rate of greenhouse gas emissions is significantly rising every year, making the country one among the top 10 emission producers in the world (Jena, 2020). Different kinds of transportation contribute to 10 % of air pollution in India, which again leads to climate change and other additional concerns like health problems.

The Indian government has been trying to promote the use of electric vehicles through various schemes such as Faster Adoption and Manufacturing of Electric Vehicles (FAME). This scheme provides financial subsidies, tax benefits, and exemptions from road taxes. These incentives are expected to boost consumers' propensity to purchase. The government incentives also include tax benefits and freedom from road taxes.

By 2030, India aims for 30% of private cars, 40% of buses, and 80% of two-wheelers and three-wheelers to be electric. The country's objective is to reduce overall emissions by 45% by 2030. Currently, electric vehicles make up about 50% of 3wheelers, about 5% of 2wheelers, and 2% of cars bought in 2024 (Wells, 2025).India has now come up with a new policy for electric vehicles, called the Scheme for Manufacturing of Electric Cars policy, which lowers import taxes on 8,000 electric cars per annum for automotive manufacturers who can invest at least $500 million in manufacturing in India that uses 50% of components sourced locally. This policy is leading local manufacturers like Tata Motors and Mahindra to be significant players now and in the future (Wells, 2025).

All the same, the most important motivator for purchase of electric vehicles still comprises the Indian consumers' growing desire, overcoming barriers, to contribute towards reducing the emission of greenhouse gases, thereby addressing climate change. (Jena, 2020).

9.4.6 Concerns over electric vehicles

In India are often related to the higher initial cost of purchase. However, increasingly, the consumers in India are realising that running costs and maintenance costs are much lower in electric vehicles as compared to traditional ones (Jena, 2020).

Also, the running of electric vehicles is much smoother and quieter, which Indian consumers find quite welcome.

To overcome the consumer perception of the high initial/purchase cost of electric vehicles, the state governments, as well as the central government, are offering incentives in various schemes. One such scheme, mentioned earlier, is called Faster Adoption and Manufacturing of Hybrid and Electric Vehicles (FAME).

However, in addition to motivators, there are also detractors, which include high cost of purchase, lack of good charging stations, charging time limitations, limitations on range of driving, high cost of replacing the battery and unsure resale value and so on.

At the current time, there is not too much awareness of climate change and the role of electric vehicles amongst the vast, not-so-educated public in the country. These obstacles and lack of awareness still need to be addressed for larger ownership of electric vehicles. (Singh et. al, 2021)

9.4.7 Opportunities of Electric Vehicles in India

India happens to be the most populated country in the world, with a population of over 140 million. Here, fuel quality has been poor, and vehicles on the road are often old and not having adequate maintenance. The road conditions are generally quite poor, and there is huge traffic congestion. So, air pollution has been a huge health threat in our country. Also, these old vehicles emit more carbon dioxide and other greenhouse gases, causing global warming and climate change.

In this situation, electric vehicles can prove to be a highly welcome initiative because of their low emission levels of greenhouse gases and also lowering air pollution.

(Singh et al, 2021)

Thus, the growth of the EV market in India is expected to be exponential, with the current market valued at around USD 5 billion, and is expected to grow to a great range of USD 47 billion by 2026. The government is promoting EV adoption with various schemes via subsidiaries and tax cuts. Various EV charging stations are being set up, creating a whole new ecosystem around the sector. The 2-wheeler EV sector has seen great response in terms of adoption and sales, proving that EVs will be a success in India. (Singh et al, 2021)

EVs are the future for another reason... maybe a more significant one. In recent times, the world has been concerned about sustainability. And EV has been at the centre of sustainability. All the major players in the Automobile Industry are moving in this direction. It's about the triple bottom line in the new age. It is not just about profits but the 3 Ps. That is People, Planet and Profits. Everybody wants to be more conscious about their carbon footprint and wants to contribute to sustainability. That is the main reason for the shift towards EVs. Governments all over the world are also promoting EVs through various schemes.

All the same, EVs have some major challenges too, as mentioned earlier. These comprise Initial Cost, Maintained Cost, Battery issues, Average speed, Driving range, charging issues, etc. These challenges still need to be addressed and overcome.

In the industry, there is a huge opportunity and much business coming in the way of EVs. There is tremendous scope for research coming in too in this area. Even now, one does see much interest and development coming from this industry. India's EV market is dominated by Tata Motors (passenger cars), Ola Electric, TVS Motor, and Ather Energy (two-wheelers). Other major players include Mahindra, Bajaj Auto, Hero Electric, and Hyundai. However, there are many other international players too who are planning to enter the Indian market.

As mentioned above, there has been much research to assess the consumer perceptions of the various characteristics of electric vehicles. However, the target population considered in such research has been various segments, such as residents in a certain area, student population in a certain area in the country or even combinations of different segments. considered together.

9.4.8 Findings of Empirical research on the perception of corporate managers on electric vehicles.

Now consider the findings of a research carried out to determine perceptions of corporate managers on electric vehicles in India. Using a text mining approach on a sample of text data, NLP (natural language processing) was used to set up frequency distributions and a word cloud, which brings out consumer perceptions of predominant features of electric vehicles which exist in the mind of the corporate consumer.

The research comes out with the finding that the corporate managers consider the capability of electric vehicles to address climate change as the most predominant feature leading to the propensity to purchase. This is the feature which would lead to a purchase decision.

The research was conducted on a target market of corporate managers because corporate managers throughout the industry have been effective in planning and carrying out successful strategic projects to bring about efficiency from their organisational perspective. It is expected and hoped that if they take it upon themselves to address the climate change impacts by promoting electric vehicles, they can be equally effective here too, this time with the goal of enhancing resilience and minimising vulnerability. (http://unfccc.int/adaptation/workstreams/nairobi_work_programme/items/4623.php).

Thus, the role of practising corporate managers with respect to electric vehicles would be twofold. On one hand, they would constitute the customer base for purchasing electric vehicles for their personal use. On the other hand, they could be the prime movers to promote electric vehicles in the country, with the objective of addressing climate change. But of course, this could only happen if they are convinced that it is possible to address climate change through use of electric vehicles because they do minimise emission of greenhouse gases and thereby address climate change. (Rao, 2020)

9.4.9 Some details of the empirical research to determine corporate managers' perceptions on the use of electric vehicles in the Indian market

A text analysis approach

In this empirical research, the objective was to gather consumer perceptions on the part of corporate managers on electric vehicles with the use of responses not in the form of numerical assessments on various concerns, but their concerns written down in a small Word file. Thus, the Target market being considered here is the population of practising corporate managers from whom a sample was considered.

As part of the research, this group of managers were requested to give their views on why they would purchase/promote electric vehicles. Their views which emerged were collated on the following features.

9.4.10 Corporate Managers' Views/Perceptions on Why They Might Purchase/Promote EVs,

Incentives:

Perception that EVs address climate change. Electric Vehicles (EVs) have emerged as a major solution in the initiatives to address climate change, a means to lower greenhouse gas emissions and lesser reliance on fossil fuels. While greatly praised for their potential environmental benefits,

EVs also present challenges, particularly in battery manufacturing, resource extraction, and energy usage.

Perception of acceptable driving cycles

This concern relates to how the vehicle would be driven... whether it will be driven on urban roads with many stops or on long distances with few stops. This is an important concern because it determines the configuration and battery pack.

Perception of the acceptable maximal speed

This forms a very important consideration for Indian consumers. The maximum speed can be anywhere up to 160 km per hour.

Perception of good maximal power

This pertains to how quickly the vehicle attains and maintains a constant speed.

Perception of public charging infrastructure

This concern relates to the availability of charging stations in the major cities.

Perception of a good driving range

Driving range is the maximum distance an electric vehicle can travel after a full charge. This can vary from 200 km to 500 km.

Perception of high initial cost, price

This forms a very major concern because often, the public perception is that electric vehicles have a high initial cost.

Perception of maintenance cost

The maintenance cost is comparatively low, which is a major reason why consumers prefer electric vehicles.

And so on.

Upon running the NLP, the research came out with the following bar plot.

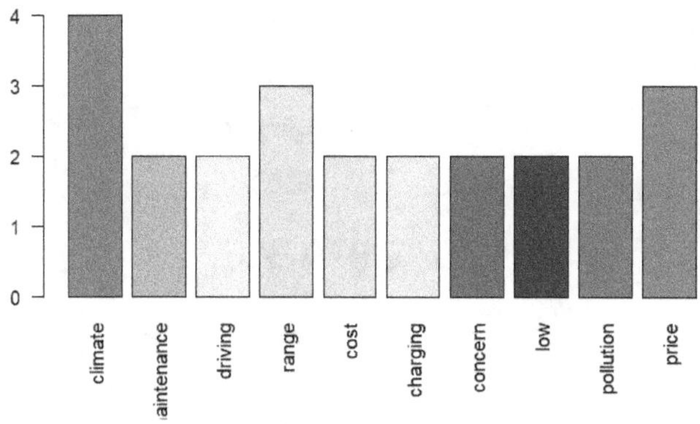

Figure 9.4.1 Bar plot for different aspects of electric vehicles, as obtained from text data.

The bar graph has the tallest bars for climate change, driving range and price. This brings out the predominant features in the mind of the corporate manager, which include the benefits of addressing climate change, an acceptable driving range, but also a high price. While climate change is the biggest incentive, the concerns of driving range and price would still need to be overcome.

All the same, most of the corporate managers emphasised that the reason for them to support the manufacture and sale of electric vehicles in India would certainly be the fact that these vehicles address climate change by reducing the generation of greenhouse gases and other emissions in the country.

Next, the package ggplot2 was used to generate the word cloud for the datafile used in the research.

The word cloud brings out that the most predominant perception is that climate change can be addressed using electric vehicles. The driving range, charging concerns, high price and other issues associated with electric vehicles still exist and would need to be overcome.

Figure 9.4.2 Word cloud for different words as obtained from text data for corporate managers.

(In the word cloud, there is no labelling of the axes. One just looks at the words which emerge **as bold and larger in size**. These words are predominant in the text.)

The word cloud brings out that the primary perception of electric vehicles is that they address climate change, which leads to the purchase of such vehicles. All the same, driving range and price concerns also emerge as highly important, though not as predominant, in the mind of the consumer. Hence, addressing climate change would be the biggest incentive for corporate managers to go for electric vehicles, while concerns over the price and driving range need to be overcome.

Conclusion and Road Ahead

The text mining research brought out the fact that the most predominant incentive for the purchase of electric vehicles in India comprises the aspect that they address climate change concerns. The next important desirable aspect is the acceptable driving range. The price certainly is still a concern. Also of concern, though not as predominant, are the availability of charging stations and the initial cost.

Given these findings, the country should spread awareness of the perils of climate change and urge corporate managers to address climate change by going for electric vehicles, thereby reducing the generation of greenhouse gases. They could also promote electric vehicles to others in their target audience, using the climate change benefit as something to go for. For this, some awareness building may be necessary through organising more and more conferences and promoting research on such topics. The managers could also advise Business Schools to include climate change courses in the course curriculum of educational institutions.

All the same, to see EVs in the Indian market, one needs to increase the awareness of Indian managers to the different aspects of EVs, actually addressing climate change, and also addressing their worries and concerns about using these vehicles. This need to determine what characteristics are at the top of the mind of the consumer, necessitated the above research, which was conducted using a text analysis approach.

To help and offer more incentives to prospective consumers, there must be continuous endeavour on the part of sellers of electric vehicles to offer discounts and lower the price of such vehicles. Also, tax incentives and price rebates offered by the government would be highly welcome.

Pointers to enhance the sale of electric vehicles in India

Based on the research findings by bar plot as well as by word cloud, the most predominant characteristic of electric vehicles in the mind of corporate managers is climate change… addressing impacts of climate change by using vehicles with reduced emissions of greenhouse gases.

This should lead the promoters of electric vehicles in India to set up campaigns highlighting the fact that if our country would like to address climate change, a very effective strategy could be to use electric vehicles. This would lead to enhanced product visibility, highlighting climate change.

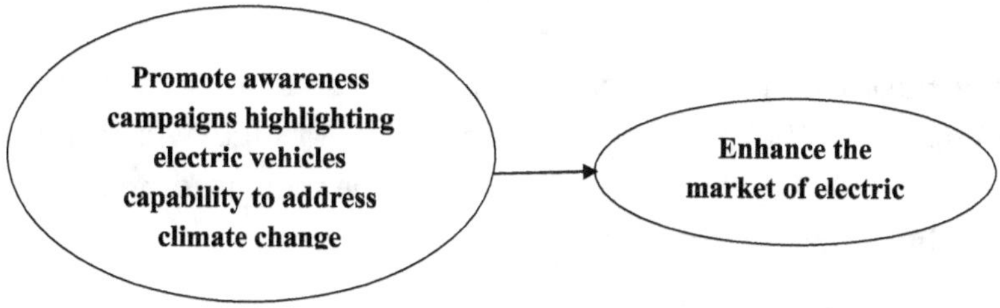

Figure 9.4.3 Market strategy to enhance the market for electric vehicles to corporate managers, predominantly highlighting climate change.

Thus, Indian corporate managers who perceive the usefulness of electric vehicles as addressing climate change could be motivated to go for electric vehicles for this attribute most of all. They could also use this finding to promote the sale of electric vehicles in the country. The automakers/manufacturers of electric vehicles could remember to include this aspect in their marketing efforts to this target market.

Pointers to market electric vehicles to the target audience of corporate managers.

The research presented above brings out the propensity of corporate managers, who constitute the target audience, towards going for electric vehicles because they address climate change. This is totally in agreement with the government's shift towards sustainable transportation which is gaining momentum. Also, all over the world, the emphasis towards Zero Emission Vehicle (ZEV) mandates in different countries has set an ambitious target, requiring manufacturers to achieve a large target of electric car sales.

Thus, the electric vehicle promoters should keep reiterating and emphasising this aspect whenever promoting electric vehicles to corporate managers.

All the same, to address the concern of price and driving range, promoters could collaborate with manufacturers to create special promotions, discounts and exclusive financing plans for zero-emission car buyers, attractive incentives and other financing options. Also, these incentives should be clearly communicated through the marketing channels, ensuring that target audiences are aware of the financial benefits of choosing an electric vehicle over regular cars. (Bott, Henry, 2024).

For the concern over driving range, it is a fact that electric vehicles (EVs) have significantly improved in terms of range over recent years. (Coltura, 2025). However, the driving can be increased by drawing up some advice for the driver, such as:

1. Timing the Charging
2. Driving Smoothly at Lower Speeds range

3. Maximising Regenerative Braking
4. Travelling Light
5. Taking Care of Tyres
6. Using Heater and AC Sparingly

(Coltura, 2025).

Additionally, the driver can adjust the driving style by accelerating and braking smoothly, using regenerative braking effectively, and reducing speed, especially on highways.

The electric vehicle is different when compared to a traditional car, but the person driving it gets used to it and overcomes the initial hurdle by slowly getting used to it.

9.5 Case 5: Exploring how Corporate Initiatives can address climate change in the Hospitality Industry

9.5.1 The hospitality industry is a broad, service-oriented sector focused on providing great customer experiences related to leisure, accommodation, and food/beverage. It includes hotels, restaurants, tourism, event planning, and entertainment, aiming to make guests feel welcomed and comfortable. Key sectors include lodging, food and beverage, travel/tourism, and recreation.

In the case of the hospitality sector, climate change is already affecting the operations in a large way. Extreme weather is increasing the cost of operations and reducing the number of tourists visiting certain destinations, while local and national environmental policies and penalties are being introduced in cities and countries around the world. Extreme weather causing floods is also causing major damage to hotels, resorts and other tourist destinations all around the Earth.

The hotel industry is a key sector within the broader hospitality industry, focusing on providing overnight accommodation and related services, such as dining, event hosting, and guest services. It is highly influenced by tourism and business travel, with current trends focusing on technology (AI), sustainability, and personalised luxury experiences, particularly in a rebounding market following pandemic-related disruptions.

This hotel sector, as part of the hospitality sector, accounts for around 1% of global carbon emissions, and this is predicted to increase. Hospitality, like other industries, has a responsibility to manage its impact on our planet, thereby reducing its carbon footprint. However, the most important aspect of climate change affecting the hotel industry depends on the way this industry is facing or adapting to the impacts of climate change, creating huge barriers to its smooth operation.

(https://sustainablehospitalityalliance.org/our-work/climate-action/)

Climate change is a pressing global concern demanding immediate attention and collaborative action across all sectors, including the hospitality industry. As the world experiences unprecedented shifts in weather patterns, rising temperatures and extreme weather events, the impact of climate change on the hospitality sector has become increasingly evident (Ben Youssef and Zeqiri 2022). The hospitality industry, which includes hotels, resorts, restaurants, and other accommodation facilities, plays a significant role in global carbon emissions due to its **energy-intensive operations** and reliance on natural resources. All the same, the way this industry is trying to adapt to climate change impacts is also exemplary.

Climate change impacts the hotel industry through increased operational costs, infrastructure damage from extreme weather, and shifting traveller preferences toward sustainability. Rising sea levels threaten coastal properties, including resorts and tourist destinations, while warmer temperatures affect seasonal, ski, and nature-based tourism. The industry must adapt by enhancing resilience, reducing carbon footprints, and addressing water scarcity.

Thus, in this industry, mitigation, reducing the generation of greenhouse gases, as well as adaptation, dealing with the impact of climate change, both need to be looked at in a significant way.

The hotel industry's energy consumption has become a focal point in global efforts to combat climate change and promote sustainability. The sector has a significant environmental footprint, with a stark contrast between the energy costs of a single hotel room and an average residential home, highlighting the significant energy consumption of the hotel industry. This significant disparity brings out the energy-intensive nature of hotel operations, which include various amenities, lighting, heating, cooling, and other services to meet the comfort and needs of guests. It is therefore increasingly important for hotels to adopt energy-efficient practices and sustainable measures to reduce their environmental impact and help mitigate climate change.

However, the impact of climate change on hotels extends beyond energy consumption, encompassing economic consequences as well. A certain research has delved into consumer perception points of view regarding environmentally friendly hotels as a strategic policy (e.g., Kim et al. 2019). For instance, ecologically innovative (green) hotels are now expected to set the benchmark in protecting the environment and mitigating human health hazards. The need for many such studies arises from the recognition that eco-innovative (green) hotels not only need to be established and promoted but also accepted and adopted by guests (Sharma and Chen 2023).

9.5.2 Corporate Initiatives to Address Climate Change Impacts

The hospitality industry, including hotels and resorts, can take many initiatives to reduce its impact on climate change and the environment. These initiatives can include:

9.5.2.1Mitigation

The hotel industry is energy-intensive, accounting for 3% of global greenhouse gas emissions. Here are some of the ways the hotel industry can help mitigate climate change:

- **Investing in energy efficiency.** Hotels can save energy with energy-efficient appliances, lighting, and HVAC systems. They can also improve the insulation of their buildings to reduce heat loss.
- **Using renewable energy.** Hotels can switch to renewable energy sources such as solar and wind power. This will help them reduce their reliance on fossil fuels and their carbon emissions.
- **Reducing water consumption.** Hotels can save water by installing low-flow fixtures, fixing leaks, and recycling water. They can also

encourage guests to conserve water by using less water in showers and baths and reusing towels and sheets.

- **Responsible sourcing.** Hotels can carefully select suppliers based on their commitment to environmental and social responsibility, and opt for local and organic products when possible. The suppliers should use renewable energy, organic farming with reduced emissions and so on.
- **Reducing food waste** because waste food gives rise to emissions which are a major source of greenhouse gas (GHG) emissions, accounting for roughly **8–10% of global emissions**. When food decomposes in landfills, it releases methane, a potent GHG, while also wasting the energy, water, and land used in its production, transportation, and disposal.
- **Offsetting carbon emissions.** Hotels can offset their remaining carbon emissions by investing in projects that reduce greenhouse gas emissions, such as planting trees or investing in renewable energy.
- **Supporting sustainable tourism.** Hotels can promote sustainable tourism by educating guests about the environmental impact of their travel and offering sustainable travel options.

The hospitality industry must accept the reality of climate change, adapt to environment-friendly business models, and fight the tide of ever-evolving challenges at the same time. ***The cost of mitigating the environmental risk for the hotel industry is not something to consider; it is part of the cost of doing business.***

9.5.2.2 Adaptation

When our environment changes, we must change with it. Unexpected weather events from extreme heat, wildfires, drought, hurricanes, storm surges, and deep freezes are the new normal, and we must adapt ourselves and our businesses.

There are regional concerns, and there are many practices that apply.

Hotels in coastal regions should work with regional and/or federal authorities to invest in infrastructure that can withstand extreme weather events, such as seawalls, storm surge barriers, and coastal erosion planning.

Coastal properties need strong risk management systems in place, including more expansive insurance and alternative revenue modules if unpredictable weather wipes out the high season.

Many regions are increasingly prone to drought and wildfire, causing significant losses to agriculture and to other flora and fauna that are often attractions for visitors. Destruction of nature can reduce a property's appeal and affect travellers coming for hiking, fishing, hunting, sporting events, rodeos, and other outdoor activities.

Droughts and water shortages make maintaining pools and other water features difficult. These catastrophic weather events can also increase the costs of sourcing local food and services.

Water stewardship is critical in these areas. The hotels and resorts could carefully calculate the water management with specialised tools from the Sustainable Hospitality Alliance. They should also develop drought-tolerant landscaping and invest in rainwater collection.

When flooding is increasing throughout the region, and with rapid climate changes, historic floodplain maps are no longer accurate. So the hotel would need to re-evaluate the property's flood plain position with the most current data and find out what the property needs to address.

All hotels must develop contingency plans to appeal to visitors and offset the potential loss of business in these circumstances.

9.5.3 Green Hospitality: Sustainable Practices for Hotels and Resorts

In today's environmentally-conscious world, the hospitality industry has begun to recognise the importance of adopting sustainable practices in its daily operations. Hotels and resorts are no longer just places to rest one's head; they have become essential in the global effort to reduce our ecological footprint. As travellers seek eco-friendly destinations, green hospitality has evolved from a mere trend to an integral component of the lodging experience.

Some of the best eco-friendly initiatives encompass a wide range of strategies, from energy-efficient lighting and water-saving measures to waste reduction and locally sourced food options. By embracing these innovations, establishments within the hospitality sector can not only minimise their environmental impact but also appeal to a growing clientele who prioritise eco-conscious choices.

9.5.4 The Hotel Industry in India

In India, the hotel industry is increasingly impacted by climate change—facing risks from extreme weather, water scarcity, and rising operational costs—while also being a significant source of carbon emissions. To combat this, the sector is gradually adopting sustainability measures like renewable energy (solar/wind), energy-efficient systems, water conservation, and waste management. The shift is driven by both regulatory pressures to meet India's 2070 net-zero goal and growing consumer demand for eco-friendly, sustainable travel options.

The industry in India is rapidly adopting sustainability to combat climate change, driven by resource scarcity, regulatory pressures, and eco-conscious travellers. Key initiatives include implementing renewable energy (solar/wind), water conservation, reducing single-use plastics, and adopting green building designs to achieve net-zero targets. Major hotel chains are integrating ESG frameworks, aiming to reduce carbon footprints and enhance operational efficiency through smart technologies. ESG (environmental, social and governance) covers a range of methods and measurements that assess businesses in ways other than their financial achievements and performance.

9.5.5 Key Aspects of Climate Change in the Indian Hotel Industry:

- Impact on Operations: Climate change threatens the industry through extreme weather affecting tourism, rising operational costs, and the need for stricter environmental compliance.

  Climate Change requires Green Initiatives & Sustainability in hotels.

- Environmental sustainability: Green energy initiatives in the hotel industry contribute to a reduction in greenhouse gases. Hotels that opt for renewable energy sources demonstrate their commitment to sustainability.

- Climate Change requires use of Renewable Energy, leading to increased investment in on-site solar, wind, and sustainable materials for construction.
- Climate Change requires proper Water & Waste Management. This leads to implementation of water-saving fixtures, greywater recycling, and the elimination of single-use plastics.
- Climate Change requires Energy efficiency. Beyond renewable energy, the hotel industry is increasingly adopting energy-efficient technologies. This includes LED lighting, smart thermostats, and energy-efficient HVAC systems. These measures reduce energy consumption, further cutting operational costs.

To achieve energy efficiency, hotels are adopting LED lighting, smart thermostats, and efficient HVAC systems.

9.5.6 The key aspects of HVAC are:

Indian hotels are slowly using this system and achieving Comfort & Safety:

This provides a comfortable, safe, and productive indoor environment regardless of the season. In HVAC, the following systems are used:

- Energy Efficiency: Modern, properly designed systems reduce energy consumption and operational costs.
- Indoor Air Quality (IAQ): Removes pollutants and controls humidity to prevent mould and bacteria growth.
- Ventilation: This is the process of exchanging or replacing air within a space to improve air quality by removing moisture, smoke, heat, dust, air-borne bacteria, and carbon dioxide.
- Air Conditioning: Units lower indoor temperatures and reduce humidity by forcing warm air over cold, refrigerant-filled coils.
- Components: Common components include air handlers, thermostats, air cleaners, humidifiers, and dehumidifiers. Major

types include split systems (most common), ductless systems, and packaged systems.

- Functionality: HVAC systems are designed based on thermodynamics, fluid mechanics, and heat transfer principles to maximise energy efficiency and sustainability.

9.5.7 Trends and Commitment

- Net-Zero Goals: The industry is working toward ambitious targets, including reducing emissions per room by 66% by 2030 and 90% by 2050.
- Corporate Commitment: Major hotel groups in India (e.g., IHCL) are integrating Environmental, Social, and Governance (ESG) frameworks, such as the "Paathya" initiative, which focuses on waste reduction and water reuse.
- Industry Trends: As eco-friendly travel grows, hotels are focusing on reducing their environmental impact to remain competitive and appeal to environmentally-conscious travellers.
- Green Hospitality The key principle of green hospitality is to create a comfortable and luxurious experience for guests while minimising the environmental impact of hotel operations. This involves adopting sustainable practices, investing in eco-friendly technologies, and promoting awareness of environmental issues.

Prioritising sustainability can be a great way for hotels and resorts to demonstrate their commitment to preserving our planet. When conducted the right way, it can also attract an increasing number of environmentally-conscious travellers.

Green hospitality extends beyond the physical property and includes educating guests about the importance of sustainable practices, engaging with local communities, and supporting environmentally responsible initiatives. This holistic approach ensures that hotels and resorts contribute positively to both the environment and society as a whole.

9.5.8 Key Trends in Indian Hotels relating to Climate Change

* Dual Role: Hotels are both significant contributors to and victims of climate change. They contribute through high energy/water consumption and waste, while also suffering from, for example, extreme weather events like flooding in Kerala.

- Sustainability Focus: To achieve net-zero, hotels are shifting toward renewable energy, using energy-efficient HVAC systems, and installing smart sensors to manage energy usage.
- Regulatory & Consumer Pressure: Increasing regulations, along with a demand for eco-friendly experiences, are forcing hotels—especially luxury ones—to adopt sustainable practices.
- Industry Examples: IHCL has launched "Paathya" to target 100% wastewater reuse, elimination of single-use plastic, and certification of all hotels, while other chains focus on Green Meetings (Innergise). In 2022, IHCL launched the Paathya framework to anchor its holistic response to the sustainability challenge. Derived from the Sanskrit word for path, 'paathya' means both way and solution.
- Potential Savings: 4-star and 5-star hotels in warm, humid regions have the highest potential for reducing emissions through better design and technology.

9.5.9 Strategies for a Sustainable Future in the Hotel Industry

- Green Building Design: Using eco-friendly, locally sourced materials and modular construction to minimise waste.
- Waste and Water Management: Advanced recycling and water-saving fixtures.
- Technology Adoption: Using IoT and AI for energy management and optimising guest services.
- Industry Collaboration: Sharing knowledge on best practices to accelerate the transition to net-zero.
- India's commitment at the United Nations in 2021 to meet net-zero emissions by 2070 has further emphasised the need for large hotels to go green.

- The hotel industry in India is rapidly adopting sustainable, "green" practices to combat climate change, aiming for net-zero emissions by 2030–2050 to reduce a significant amount by implementing solar power, waste management, and LEED certification to meet eco-conscious consumer demand and stricter regulations.

9.5.10 Advancements in climate change-friendly technologies in Indian hotels

Energy-Saving Technologies in Indian Hotels

As excessive energy consumption remains a significant concern in the hospitality industry, many hotels and resorts have turned to innovative technologies to reduce their carbon footprint. These include LED lighting, solar panels, and energy-efficient heating and cooling systems. Additionally, smart sensors can be utilised to adjust room temperature and lighting based on occupancy, ensuring energy is not wasted.

Thanks to investing in these advanced solutions, hotels can significantly cut down on their energy usage while still providing a comfortable and enjoyable experience for their guests. Moreover, these energy-saving measures will result in significant cost savings for the establishment over time.

Waste Management and Reduction Strategies in Indian Hotels

Effective waste management is crucial for any eco-friendly establishment. Indian hotels are trying to adopt various strategies to minimise waste production, such as implementing recycling programmes, reducing single-use plastics, and offering reusable alternatives for items like linens and toiletries. Composting initiatives also help transform food waste into valuable resources for local agricultural projects.

Furthermore, educating staff members on proper waste disposal procedures is equally important in ensuring these initiatives succeed. Food waste, in particular, is a prevalent issue within the hospitality industry, and hotels can combat this by implementing portion control measures in restaurants or offering customisable meal options. Donating surplus food to local food banks or charities is another excellent way to minimise food waste while supporting the community.

Sustainable Sourcing and Procurement

Responsible sourcing and procurement practices can contribute significantly to a hotel's sustainability efforts. This involves carefully selecting suppliers based on their commitment to environmental and social responsibility, as well as opting for local and organic products whenever possible.

When prioritising eco-friendly materials and supplies, hotels can reduce their carbon footprint, support local economies, and promote healthier environments for both guests and staff. Moreover, collaborating with suppliers to reduce packaging waste and streamline delivery processes can further enhance a hotel's green credentials.

Water Conservation Measures

Water conservation is another vital aspect of green hospitality. Indian hotels are adopting various measures to conserve water, such as installing low-flow fixtures, implementing greywater recycling systems, and utilising drought-resistant landscaping. These initiatives not only preserve precious water resources but also result in significant cost savings for the establishment.

Educating guests about the importance of water conservation and encouraging responsible usage is another essential component of a hotel's water-saving strategy. For instance, promoting towel and linen reuse programmes can be an excellent way to conserve water within a hotel. By encouraging guests to use their towels and linens for more than one day, hotels can significantly reduce the amount of water, energy, and detergent used in laundry operations.

Eco-Friendly Amenities and Services in India

Hotels and resorts are providing electric vehicle charging stations for use in hotels and resorts.

Offering eco-friendly amenities and services allows hotels to cater to environmentally-conscious travellers while minimising their impact on the environment. Examples include providing natural toiletries in biodegradable packaging, offering bicycle rentals or electric vehicle charging stations, and promoting nature-based activities that promote environmental awareness.

Establishing on-site gardens or green spaces is another approach hotels are taking to minimise their environmental impact. These areas can provide fresh produce for hotel restaurants, reduce heat island effects, and enhance air quality - all while offering an appealing atmosphere for guests to enjoy.

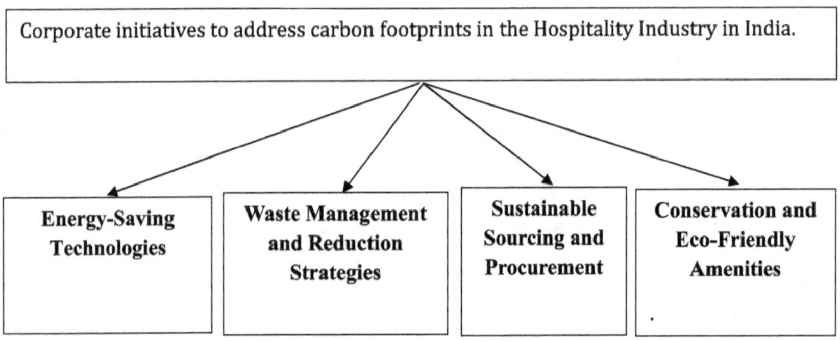

Figure 9.5 Corporate initiatives to address carbon footprints in the Hospitality Industry in India

In Conclusion

In India, green hospitality has emerged as a crucial aspect of the modern lodging experience, driven by growing environmental awareness and the demand for eco-conscious travel options.

Today, there's no doubt that by adopting sustainable practices, such as energy-saving technologies, waste management strategies, responsible sourcing, and water conservation measures, hotels and resorts can significantly reduce their environmental impact and cater to the needs of environmentally-conscious travellers.

Moreover, these establishments can enjoy cost savings, enhance their reputation, and contribute to a greener future for all. Promoting such practices enables the hospitality industry to continue driving innovation and allows it to contribute positively to both the environment and society.

9.6 Case 6: Hotel Azure, How a South-East Asian Resort Hotel addressed climate change:

How a celebrated resort hotel implemented Adaptation initiatives across the supply chain to address Climate Change.

Let us consider a much celebrated resort hotel which was established about 40 years ago on the golden coastline bordering the blue waters near an ocean. The region around the hotel has had the great opportunity to host tourists coming from all over the world for ages. The coastline was studded with coconut trees and mountain ranges, which gave it an idyllic charm. The region was discovered originally by pirates in the 16th century and converted into a bustling tourism and trading centre. With many traders visiting the place and many historic events taking place in the surrounding region, the locality became a melting pot of fascinating cultures, religion and food. This, combined with the tranquillity of the seas, provided a perfect backdrop for the hotel. The hotel experienced tremendous growth in its size, the number of customers and volume of business. Along with the growth, the hotel also received many awards and recognition from different national agencies for best in location, best in food services and best in landscape competition. All these awards and recognition demonstrated that the hotel's search for excellence in service, quality and customer satisfaction was in existence all along.

While serving as a favourite holiday resort for a huge number of individual customers with families, care was taken to provide a great outdoor experience along with exemplary ambience, excellent food and natural setting. Along with serving individual customers, the hotel also served as a business retreat and provided large convention rooms where many companies would hold their strategic planning workshops. Also, many wedding parties would arrive to hold celebrations against the wonderful ocean backdrop.

All the same, the hotel was starting to get threatened by climate change impacts almost every year.

Around the yearly monsoon season, the sea level seemed to rise more and more every year. The communities from where the hotel's staff and suppliers came had agricultural land and fruit orchards within a few kilometres of the hotel premises. They were in trouble because their houses, roadways, water supply, school buildings, electricity generation systems and even their fruit orchards would get submerged by salty sea water. Some communities staying there had agricultural land where they grew rice crop, vegetables, various green spices and fruits and supplied them to the hotel. They also maintained cattle farms and poultry farms from where the egg and meat products came. The flooding of seawater, which submerged their lands for days, enhanced the salinity in the land so much that instead of three crops, only two or one crops were being grown by the farming communities there.

Apart from the farming communities, from whom much of the hotel's daily vegetable and meat supplies came, the hotel also had seafood suppliers who would take their fishing boats out into the ocean every morning and return with fresh catch of fish, shrimps, oysters and other seafoods in the afternoon. Most of the catch was sold fresh to the hotel and used for serving food in the hotel kitchens. However, the warming sea waters, another impact of climate change, was depleting the availability of fish, so the fishermen were often not able to supply the full amount of seafood required by the hotel.

Let us now consider the **adaptation initiatives** the hotel took across the supply chain.

9.6.1 Inbound phase... putting up dikes and check dams.

To address the salinity of the agricultural land, the hotel put up 10-foot-high dikes along the seashore bordering the land where the farming communities lived. The dikes ran for several kilometres along the coast. This prevented the flooding seawater from entering the agricultural land. Also, the housing neighbourhoods, water supply stations, schools, roadways, markets, electricity generation plants and other infrastructure were protected from getting submerged.

The agricultural land also had small fresh water seasonal rivers, which flowed into the ocean but dried up during the summer. The hotel invested and put up check dams to allow for better percolation and recharge of underground water. The dams helped in putting up reservoirs for potable water for all. Existing village ponds were de-silted, deepened, and interlinked. This resulted in improved agricultural productivity. Instead of one crop per year, farmers are able to grow three. Overall, agricultural productivity increased a lot with the enhanced availability of sweet water from the interlinked rivers. The flower orchards and horticulture prospered. Also, improved quality of drinking water led to improved health status of the population. The saline water had caused kidney stones and problems with the bones - these problems declined.

To enhance the availability of fish & seafood, the hotel constructed water reservoirs where various kinds of fish were cultivated. This helped to increase the supply of fish to the hotel to address the falling levels of fish production, which had been depleted by the warming seawater.

9.6.2 Adaptation Initiatives in the Production Phase

In the context of the Resort hotel, the production phase comprised the production of the hoteling service... providing ambience and comfortable accommodation to hotel guests, excellent food, sports & gymnastic facilities, water sports activities, great outdoor space and so on.

Because of climate change, the water level was continuously rising over the years, making the high tide waves much higher and stronger than before. The high tide would come with much larger elevation and with stronger force. This caused huge cracks in the sand on the open beach area, where the hotel guests loved to walk around, making the area rather dangerous and unsuitable for walking. The area also proved to pose potential threats for families who came to the beach for swimming.

The stronger and stronger high tide waves, splashing on the beach area, weakened the ground under the hotel structure, so some parts of the hotel walls were sinking into the ground as well. The stronger waves also damaged the foundation and increased the salinity of the ground, causing the hotel gardens to deteriorate.

To address these issues, the hotel implemented various adaptation initiatives, such as repairing the cracks on a daily basis with high-potency cement. It could not build dikes along the coast in this area because the great ocean view of the hotel would get obstructed, but it did plant some coconut trees and huge orange-flowered flame trees to strengthen the ground. The continuous cementing of the cracks also helped to strengthen the ground under the beach.

9.6.3 Adaptation Initiatives in the Outbound Phase

Adaptation in **waste management**. The waste generated in the hotel was mainly kitchen waste, glass and metal containers, paper and plastic packaging materials. These were all non-hazardous because cleaning agents, laundry detergents, etc., were all non-hazardous types. The liquid waste was taken to the recycling ponds and treated with non-hazardous neutralising agents. The clear water which flowed out was used in gardening, and the solid leftover residue was combined with cement and used for making walk paths or to continuously fill up the cracks caused by the strong waves. The other waste of glass, plastic, etc., was given to waste handlers.

Green building. The hotel also followed all guidelines of green building and put up rows of clay pots lining the roof where various plants were grown, which helped absorb the sun's rays, keeping the building cool.

Outreach to build climate change awareness. There was one more initiative in the adaptation area which the hotel carried out. Among the major customers who came to the resort on a regular basis were corporations which held various events, such as planning retreats in the hotel. They brought groups of employees who stayed and attended conventions & workshops in the hotel. For these customers, the hotel offered awareness-building seminars to address climate change. Sometimes these seminars were held in the hotel premises. Sometimes, employees of the hotel who were involved in Adaptation initiatives would go and address employees of the customer company at their premises.

This initiative was given a lot of accolades in the media and brought honour as well as awards from the industry.

Just like the adaptation initiatives, the hotel also carried out many mitigation initiatives too, across the entire supply chain. For instance, the hotel cars were all powered by natural gas or were electric vehicles. The cleaning agents were specially chosen so that they did not generate emissions. All cooking was carried out using renewable fuels, and all lighting and water heating were powered by LED lights and solar water heaters.

Some awareness-building initiatives to help adaptation in Hotel Azure

The first poster refers to conservation of energy, water, etc.

The second poster refers to reducing all kinds of waste which gives rise to GHGs.

The waste mentioned in the posters also includes plastic waste, which is a major source of greenhouse gas (GHG) emissions, contributing roughly 3.4% of total global emissions. Emissions occur throughout its lifecycle, from fossil fuel extraction for production to waste management, with incineration and degradation in the environment releasing methane and ethylene.

Figure 9.6.1 Implementing Energy Conservation

Source:

(1) Greening the Supply Chain, A Guide for Asian Managers, (Rao, Purba) Sage, 2008.

(2) Towards a Green Millennium, Environmental Management Systems in South East Asia, Rao, Purba, 2001.

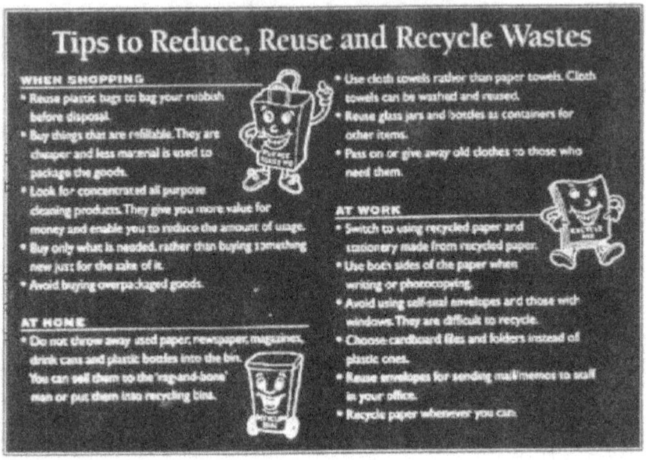

Figure 9.6.2 Reduce, Reuse and Recycle Wastes

Source:

(1) Greening the Supply Chain, A Guide for Asian Managers, (Rao, Purba) Sage, 2008.

(2) Towards a Green Millennium, Environmental Management Systems in South East Asia, Rao, Purba, 2001.

CHAPTER 10

THE ROLE OF ARTIFICIAL INTELLIGENCE IN WINNING THE BATTLE AGAINST CLIMATE CHANGE

This Chapter has been written by **Arjun Isaac Chan-Halady.**

Artificial intelligence, AI, is fundamentally reshaping how organisations manage complexity, evolving far beyond a novel technological tool to become the central nervous system of the modern, sustainable company. In the context of corporate climate action, AI acts as an unparalleled analytical engine—one capable of ingesting vast oceans of fragmented, high-velocity data, from global weather models and satellite imagery to granular supply chain logistics and real-time facility energy consumption, and translating that data into precise, predictive intelligence. For business leaders, AI is the bridge between ambitious sustainability pledges and the physical, operational reality of achieving them.

For a CEO, the imperative to embed artificial intelligence into core operations has never been more urgent. The transition to a net-zero economy and the escalating physical realities of climate change cannot be managed with static spreadsheets, siloed data, or retrospective annual reports. The sheer scale, speed, and unpredictability of climate volatility demand dynamic, real-time decision-making grounded in massive volumes of data. CEOs who attempt to tackle these compounding challenges without AI will find their organisations quickly outpaced by both environmental disruptions and technologically advanced competitors. This is no longer simply an operational upgrade—it is a strategic necessity that moves a company's posture from performative, compliance-driven sustainability to structural, profitable transformation, creating a competitive moat that protects both the planet and the bottom line.

The evidence for AI's climate potential is striking. A 2025 study by the Grantham Research Institute on Climate Change found that, if applied wisely to policy design, insight generation, and system monitoring, AI could reduce global emissions by 3.2 to 5.4 billion tonnes of CO_2-equivalent annually by 2035.) Enterprises leveraging AI for sustainability are already seeing up to an 8 times return on investment in climate risk mitigation, according to Microsoft. The question is no longer whether AI belongs in a company's climate strategy—it is how quickly leadership can act.

In this chapter, we explore how AI empowers executive leadership on three critical fronts. First, we examine how machine learning revolutionises carbon measurement, enabling precise, audit-ready tracking of Scope 1, 2, and 3 emissions without the immense manual overhead that once made comprehensive accounting impractical. Second, we explore how AI enhances corporate climate resilience through advanced predictive analytics, allowing managers to foresee physical risks—extreme weather events, agricultural shifts, resource shortages—months in advance and act before disruption strikes. Finally, we examine how AI simultaneously reduces operating costs and carbon footprints through the automation of daily operations, from smart building systems to predictive maintenance. Ultimately, artificial intelligence proves that decarbonisation and operational efficiency are not competing mandates, but parallel outcomes of an intelligently managed business.

Measuring and Auditing Carbon Footprints: Persefoni

- The carbon accounting software market tells its own story about urgency: valued at $18.56 billion in 2024, it is growing at a compound annual rate of 22.7%, on track to reach $22.77 billion in 2025 alone. (Persefoni, 2025). At the centre of this market is Persefoni, a climate management and carbon accounting platform recognised by Forrester as a leader in its category. Persefoni acts as a centralised hub for corporate carbon data, helping companies measure, report, and reduce their greenhouse gas emissions while simplifying compliance with global regulatory frameworks. Its AI automates data ingestion, detects statistical anomalies across massive datasets, and uses natural language processing to instantly map purchasing activities to the correct emission factors—eliminating manual tracking, enhancing audit-ready accuracy, and empowering organisations to transition from basic disclosure to strategic climate action.

Four Core Capabilities

- **Automatic data integration**. Through its "Integration Hub," Persefoni connects directly to a company's existing software—AWS, SAP Concur, NetSuite, utility providers—to automatically pull activity and spend data, eliminating the need to manually synchronise between disparate sources.
- **AI-powered mapping and quality control**. The platform's AI ingests natural language descriptions, such as procurement line items or utility bills, and instantly maps them to the correct emission factors. It simultaneously flags outliers, missing records, and data gaps before they become compliance liabilities.
- **Comprehensive emissions tracking**. Persefoni calculates Scope 1 (direct), Scope 2 (indirect/energy), and the notoriously difficult Scope 3 (supply chain/value chain) emissions—the category that accounts for the vast majority of most companies' climate impact.
- **Audit-ready reporting**. The platform's "Footprint Ledger" ensures every calculation is fully traceable. It automatically formats data to comply with major global regulatory frameworks, including the SEC, the EU's Corporate Sustainability Reporting Directive (CSRD), the International Sustainability Standards Board (ISSB), the Greenhouse Gas Protocol (GHGP), and California's SB 253.

From Compliance to Competitive Advantage

- Companies implementing Persefoni transition from guessing their carbon output to actively managing it. The results materialise across four dimensions. First, dramatic time and cost savings: what traditionally took months of consulting work is automated, with Persefoni Pro users able to calculate a footprint in minutes after data collection. Second, regulatory resilience: as climate disclosure laws tighten globally—from California's SB 253 to Europe's CSRD—companies eliminate the risk of non-compliance and greenwashing accusations by anchoring disclosures in verifiable, audit-ready data.

- Third, actionable decarbonisation: because the data updates near real-time rather than annually, companies can identify immediate emissions "hotspots," set science-based targets, and model reduction trajectories with precision. Fourth, supply chain competitive advantage: small and medium-sized businesses (SMBs) that calculate their emissions become highly attractive vendors to Fortune 500 companies legally mandated to clean up their Scope 3 footprints.
- The platform's reach across industries underscores its versatility. Banks, asset managers, and private equity firms use it to calculate "financed emissions"—the carbon footprint of the companies they invest in or lend money to—through full integration of the Partnership for Carbon Accounting Financials (PCAF) standards. Large corporations in manufacturing, tech, retail, and logistics use it to manage massive datasets across global facilities. For SMBs, Persefoni Pro—a free tier launched in 2024—welcomed over 3,000 users within its first six months, including universities, local governments, NGOs, and businesses worldwide, solving the critical Scope 3 "data gap" where enterprises previously had to rely on inaccurate industry averages.

Examples in Practice

Pleo (Spend Management) https://www.pleo.io/en

- To prepare for the EU's CSRD (https://csrdinstitute.eu/about), Pleo used Persefoni's AI-powered platform to calculate its baseline carbon footprint. Persehttps://csrdinstitute.eu/aboutfoni's AI transformed Pleo's spend-based data—spread across multiple databases and sources—by automatically categorising and mapping purchasing activities to accurate emission factors. This AI-driven automation drastically reduced the administrative burden on Pleo's lean sustainability team and produced the robust, audit-ready data traceability required for strict regulatory compliance, allowing the company to scale its decarbonisation strategy without scaling its headcount.

TPG (Private Equity)

- TPG (formerly Texas Pacific Group) is a leading global alternative asset management firm founded in 1992.
- Major private equity firm TPG needed to understand the climate impact of its vast investment portfolio. They deployed Persefoni Pro across their portfolio companies—ranging from a fintech firm in South Africa to a solar company in the US—giving businesses with zero prior carbon accounting experience the tools and expert support to accurately measure their Scope 1 and 2 emissions. This generated the granular, bottom-up data TPG needed to fulfil its reporting obligations to its own investors, setting a model for how private equity can drive sustainability accountability across entire portfolios.

Aleph Farms (Cellular Agriculture). Aleph Farms, a cultivated meat innovator, wanted to build a "net-zero from the start" business model rather than retrofit sustainability later. Using Persefoni, they tracked emissions at a highly granular level across their entire supply chain—Scopes 1, 2, and 3—from the earliest stages of operations. This foresight allowed them to align aggressive growth plans with their 2030 net-zero targets without the costly corrections that haunt companies that delay this work.

Enhancing Climate Resilience Through Prediction: ClimateAi

- Climate events are becoming increasingly unpredictable, and what were once "once-in-100-years" events are now occurring with alarming regularity. Some climate risks that were readily insurable a decade ago are today either prohibitively expensive to insure or effectively uninsurable altogether. Climate resilience has moved from the periphery to the centre of the CEO agenda.

- **ClimateAi is an enterprise climate resilience platform** that uses "physics-driven AI" to forecast hyper-local weather and climate impacts up to six months in advance. By applying machine learning to combine global climate simulations with satellite and weather station data, ClimateAi translates massive datasets into precise, actionable business metrics—allowing companies to proactively reroute supply chains, adjust agricultural planning, and protect their bottom lines from weather-driven disruptions before they strike.
- Research from the Frontiers in Climate journal confirms AI's structural advantage here: AI-based predictive models enable organisations to anticipate droughts or pest outbreaks to safeguard crop yields, design smarter infrastructure networks that continue functioning during extreme events, and accelerate damage assessment through satellite imagery pattern recognition—capabilities that legacy tools simply cannot replicate.

Here are five ways a CEO can use ClimateAi to align operations with the real risks posed by a changing climate.

1. De-risking Supply Chains to Avoid Major Losses.

   ClimateAi provides hyper-local (1km resolution) forecasting up to six months ahead. Rather than reacting to a weather disaster after it disrupts operations, a CEO can use these insights to foresee supply shocks—flagging a high probability of severe drought, unseasonal frost, or flooding in a key sourcing region—and pre-emptively reroute logistics, adjust harvest schedules, or build inventory buffers. This significantly reduces the millions of dollars typically lost to wasted goods, emergency logistics premiums, and stalled operations.

2. Strategic Procurement and Price Hedging.

   For companies relying on raw materials—especially in food, beverage, consumer goods, and agriculture—weather directly drives commodity price volatility. ClimateAi's models predict how changing weather will impact specific crop yields and material availability across regions. Procurement teams can use this foresight to lock in spot prices before a regional shortage drives costs up, and identify "climate analogues" – new geographic regions where essential materials will thrive as conditions shift. This protects profit margins from sudden price spikes and ensures supply continuity when less-prepared competitors face empty warehouses.

3. Quantifying the ROI of Adaptation Investments.

   Investing in physical climate adaptation – installing new irrigation systems, relocating facilities, changing seed varieties – is often stalled because the financial return is difficult to prove. ClimateAi allows executives to run financial simulations on specific adaptation measures, calculating the exact return of investing in water-efficiency technology versus the cost of inaction. This ensures capital flows only to resilience projects with proven, positive financial returns, preventing both over-investment and negligent under-investment.

4. Capturing Market Share During Disruptions.

   Climate resilience is not only a defensive play – it is a source of distinct competitive advantage. By seeing weather disruptions before competitors do, a CEO can position the company to meet sudden shifts in consumer demand. A roofing materials manufacturer used ClimateAi to anticipate elevated hurricane risks in Florida ahead of Hurricane Ian, well before traditional forecasts signalled danger. By ramping up production and securing raw materials early, the company met the massive post-storm spike in demand faster than its rivals – resulting in an additional $15 million in sales.

5. Standardising Enterprise Risk Management.

   Climate volatility makes rigid, annual planning obsolete. ClimateAi provides "Adaptation Playbooks" that set specific operational thresholds across an organisation, aligning departments – operations, finance, procurement – around a single source of truth for climate data. If a specific weather threshold is breached in the forecast, the system automatically alerts teams to trigger contingency plans, ensuring the entire executive team responds to climate threats swiftly, efficiently, and in concert.

Reducing Operating Costs While Reducing Climate Impact: BrainBox AI

- Electricity costs—particularly air conditioning—represent one of the largest line items in any commercial real estate budget, especially in warmer climates. BrainBox AI addresses this directly by doing what human operators and static schedules cannot: continuously analysing thousands of dynamic variables—granular weather forecasts, occupancy trends, and electricity tariffs—to optimise HVAC systems in real time. Managers connect this autonomous platform to their existing building management systems without expensive, disruptive hardware retrofits. Once integrated, the AI pre-emptively adjusts heating and cooling, zone by zone, resulting in an automated building that slashes total HVAC energy costs by up to 25%, reduces carbon emissions by up to 40%, and improves occupant comfort by 60%. The platform is now deployed in more than 15,000 buildings across 70 cities worldwide, including 60 major hypermarkets in the United States.
- The real-world results are equally compelling. At 45 Broadway in Manhattan, implementing BrainBox AI reduced HVAC energy consumption by 15.8%, saving over $42,000 annually and mitigating 37 metric tonnes of CO_2-equivalent in under a year—with minimal installation friction, requiring only a software integration.

Three Strategic Use Cases

- Autonomous HVAC Energy Optimisation.

  (HVAC stands for Heating, Ventilation, and Air Conditioning, referring to the technology and systems used to regulate temperature, humidity, and air quality in residential, commercial, and industrial buildings.)
- **BrainBox**, AI's deep learning algorithms, continuously ingest thousands of data points—weather forecasts, real-time occupancy

levels, and utility tariff structures—to predict a building's thermal behaviour zone by zone, making micro-adjustments every few minutes. This proactive approach eliminates energy waste and removes the need for manual human intervention, while also extending the operational lifespan of HVAC equipment.

- Generative AI for Predictive Maintenance.

  Through its generative AI assistant, ARIA, BrainBox AI transforms how facility managers interact with their buildings. Rather than sifting through thousands of system alerts, managers can ask ARIA—via text or voice—to diagnose an issue, such as why a specific floor is too cold. ARIA instantly analyses historical equipment data and performance trends to perform root-cause analysis and recommend maintenance actions before full system failures occur, dramatically cutting both equipment downtime and the number of required contractor site visits.

- Portfolio-Wide Decarbonisation and ESG Reporting.

  For retail chains and large real estate portfolios, BrainBox AI serves as a centralised decarbonisation engine. The platform continuously tracks energy consumption across hundreds of distributed locations, measures Scope 1 and 2 greenhouse gas emissions, and benchmarks each building's energy usage intensity against comparable properties. Facility managers can generate standardised carbon emission reports for compliance purposes or identify opportunities for voluntary carbon offsets—creating a transparent, automated pathway to net-zero at portfolio scale.

The Road Ahead: Agentic AI and the Next Frontier

- The tools explored in this chapter represent the leading edge of AI's current application to climate action. But the horizon is shifting quickly. In 2026, the industry has accelerated its move toward agentic AI—systems that don't just analyse data or respond to queries, but autonomously plan, execute, and coordinate multi-step tasks, including managing other AI agents. Gartner identifies agentic AI as the top strategic technology trend for 2025, predicting that 33% of enterprise software applications will include agentic AI by 2028, up from less than 1% in 2024. Separately, Gartner projects that 40% of enterprise applications will embed task-specific AI agents by 2026.

- The implications for corporate climate action are profound. At Schneider Electric, for example, an Emissions Measurement Data Transformation Agent already transforms messy, unstructured client data into carbon-calculator-ready formats in seconds—a process that previously took hours or days. Multi-agent systems will soon allow companies to deploy coordinated networks of AI agents that simultaneously optimise energy consumption, negotiate supply chain reroutes, generate regulatory disclosures, and flag emerging climate risks—all without human prompting. PwC research shows that enterprises deploying optimised AI systems are already achieving results that include conserving energy usage by 53%, cutting operational carbon emissions by 38%, and reducing costs by 61%. (Chapman, Tom, 2026).
- The convergence of agentic AI with the platforms described in this chapter—carbon accounting, climate forecasting, building optimisation—points toward a future where a company's entire sustainability operation is not just informed by AI, but actively orchestrated by it. As long as there is continued focus on reducing the impacts of climate change, agentic AI will enable companies to accelerate their efforts and achieve impact at a scale that was previously unimaginable. For the CEO reading this book, the message is clear: the tools exist, the evidence is in, and the competitive gap between those who act and those who wait is widening every quarter.

REFERENCES FOR CHAPTER 10

- Amazon Web Services. "BrainBox AI Makes Buildings Greener and Smarter with AWS Generative AI." *AWS Blog*, 11 Sept. 2024,

  https://aws.amazon.com/blogs/aws-insights/brainbox-ai-makes-buildings-greener-and-smarter-with-aws-generative-ai/.

- Ayadi, R, Yeganeh Forouheshfar2* and Omid Moghadas, (2025) Enhancing system resilience to climate change through artificial intelligence: a systematic literature review, Frontiers in Climate Journal.

- BCG and CO2 AI. "Four Out of Five Companies Report Financial Gains from Climate Action While 70% Are Maintaining or Increasing Overall Investment." *Boston Consulting Group*, 15 Sept. 2025,

  https://www.bcg.com/press/16september2025-financial-gains-climate-action.

- BrainBox AI. "2024 in Review: AI's Influence on Building Management (and Beyond)." *BrainBox AI*, 16 Jan. 2025,

  https://brainboxai.com/en/articles/2024-in-review-ais-influence-on-building-management-and-beyond.

- BrainBox AI. "BrainBox AI to Cut Commercial Real Estate Emissions by up to 40%." *Hello Future*, Orange, 28 Jan. 2025,

  https://hellofuture.orange.com/en/brainbox-ai-to-cut-commercial-real-estate-emissions-by-up-to-40/.

- Carbon Direct. "Top 5 Climate Topics to Watch in 2025 (So Far)." *Carbon Direct Insights*, 25 Feb. 2026,

  https://www.carbon-direct.com/insights/top-5-climate-topics-to-watch-in-2025-(so-far).[5]

- Chapman, Tom (2026) https://www.weforum.org/stories/authors/tom-chapman
- DeepMind. "DeepMind AI Reduces Google Data Centre Cooling Bill by 40%." *DeepMind Blog*, 19 July 2016,

  https://deepmind.google/blog/deepmind-ai-reduces-google-data-centre-cooling-bill-by-40/.
- Ernst & Young Global Limited. "2025 EY Global Climate Action Barometer." *EY*, 4 Nov. 2025,

  https://www.ey.com/en_gl/insights/climate-change-sustainability-services/climate-action-barometer-survey.
- Fujitsu. "Agentic AI: A Blueprint for Sustainable Organizational Transformation." *Fujitsu Global Blog*, 2 June 2025,

  https://corporate-blog.global.fujitsu.com/fgb/2025-06-03/01/.
- McKinsey & Company. "Cloud-Powered Technologies for Sustainability." *McKinsey & Company*, 8 Nov. 2023,

  https://www.mckinsey.com/capabilities/tech-and-ai/our-insights/cloud-powered-technologies-for-sustainability.
- Persefoni. "AI-Powered Climate Management and Reporting – Persefoni AI." *Persefoni*, 31 Dec. 2024,

  https://www.persefoni.com/persefoniai.
- Persefoni. "Persefoni 2025 Year in Review." *Persefoni*, 17 Dec. 2025,

  https://www.persefoni.com/blog/persefoni-2025-year-in-review.
- Phyniks. "AI in Sustainability: 7 Use Cases & AI Agents Transforming 2026." *Phyniks*, 30 Oct. 2025,

  https://phyniks.com/blog/ai-in-sustainability-7-use-cases-ai-agents-transforming-2026.
- Schneider Electric. "Climate Week NYC 2025 Reflections: Powering the Future with AI and Digital." *Perspectives by Schneider Electric*, 15 Oct. 2025,

  https://perspectives.se.com/blog-stream/climate-week-2025-recap.

- Artificial Intelligence's Energy Paradox: Balancing Challenges and Opportunities in the Race to Net Zero." *World Economic Forum*, 2025,

  https://reports.weforum.org/docs/WEF_Artificial_Intelligences_Energy_Paradox_2025.pdf.
- "How Tech Giants Can Scale AI without Breaking Their Climate Commitments." *World Economic Forum*, 21 Oct. 2025,

  https://www.weforum.org/stories/2025/10/ai-carbon-debt-carbon-removal/.
- "How AI Is Making Buildings More Energy-Efficient." *Time*, 10 Dec. 2024,

  https://time.com/7201501/ai-buildings-energy-efficiency/.

REFERENCES FOR CHAPTERS 1 – 9

Ahmed, Saleem (2024), The Relationship Between Climate Change and Promoting Sustainability in the Semiconductor Manufacturing Sector, TECHNICAL TEXTILES TODAY

Alexandratos, A and Bruinsma, J. (2012), World Agriculture Towards, 2030/2050, ESA Working Paper No. 12-03, FAO.

Beamon, Benita M. (2006), Sustainability and the Future of Supply Chain Management Operations and Supply Chain Management, Vol. 1, No. 1, May 2008, pp. 4-18

Berggren, A, Jansson, A. Low, M. (2019), Approaching Ecological Sustainability in the Emerging Insects-as-Food Industry, Trends in Ecology & Evolution

Biagini, B., & Miller, A. (2013). Engaging the private sector in adaptation to climate change in developing countries: Importance, status and challenges. *Climate and Development, 5*(3), 242-252.

Boojh, R. (2009). *Environment and health issues related to climate change.* National Symposium on Environmental Threat to Human Health in 21st Century, MMV, Banaras Hindu University. January, 16-17.

Burch and Robinson 2007, Dowlatabadi, 2007, Goklany 2007, Jones, Dettmann, Park, Rogers, and White 2007, Klein et al. 2007, and Swart and Raes, 2007).

Burch, S., & Robinson, J. (2007). A framework for explaining the links between capacity and action in response to global climate change. *Climate Policy,* 7(4), 304-316.

CEO Water Mandate (2014) Corporate Water Disclosure Guidelines Toward a Common Approach to Reporting Water Issues.

Ceres, 2009, Clouds and the Earth's Radiant Energy System (CERES) project, measures Earth's radiation budget,

Ecosystem, mitigation and adaptation https://www.cifor-icraf.org/publications/pdf_files/Books/BLocatelli160138.pdf, https://19january2017snapshot.epa.gov/climate-impacts/climate-impacts-ecosystems_.html

https://wotr.org/2022/02/23/what-is-ecosystem-based-adaptation-and-how-can-it-help-india-fight-climate-change/

Deshingkar, P., Akter, A. (2008) http://xue Migration and Human Development in India https://ideas.repec.org/p/hdr/papers/hdrp-2009-13.html

Dowlatabadi, H. (2007). On integration of policies for climate and global change. *Mitigation and Adaptation Strategies for Global Change,* 12(5), 651-663.

Downing, T. E. (2001). Climate Change Vulnerability: Linking Impacts and Adaptation. Report to the Governing Council of the United Nations Environment Programme. Environmental Change Institute, Oxford, UK

Fankhauser, S., Smith, J. B., & Tol, R. S. J., (1999). Weathering climate change: some simple rules to guide adaptation decisions. Ecological Economics, 30, 67-78.

Filtho, Walter Leal, Patsy Perry*, Hilde Heim, Maria Alzira, Pimenta Dinis, Haruna Moda4, Eromose Ebhuoma and Arminda Paço, An overview of the contribution of the textiles sector to climate change, Frontiers in Environmental Science, 2022.

GEF. (2009). *Facilitating an international agreement on climate change: Adaptation to climate change.*

Goklany, I. M. (2007). Integrated strategies to reduce vulnerability and advance adaptation, mitigation, and sustainable development. *Mitigation and Adaptation Strategies for Global Change,* 12(5), 755-786.

Grail Research, 2009, Research. (2009). Water, the India Story. http://www.grailresearch.com/pdf/ ContenPodsPdf/Water- The_India_Story.pdf. accessed on October 12, 2017

Halady, I., & Rao, P. (2010). Can awareness to climate change lead to behavior change? International Journal of Climate Change, Strategies and Management, 2(1), 6-22.

Hugh, T., Burke, J. and Faures, Jean-Marc (2011), Climate Change, Water and Food Security, Food and Agricultural Organization, United Nations, Rome, Italy.

Jain, S. K. (2011). Sustainable water management in India considering likely climate and other changes. *Current Science, 102*(2), 25.

Jena, R. (2020), "An empirical case study on Indian consumers' sentiment towards electric vehicles: A big data analytics approach", Industrial Marketing Management

Volume 90, October 2020, Pages 605-616.

Jones, R. N., Dettmann, P., Park, G., Rogers, M., & White, T. (2007). The relationship between adaptation and mitigation in managing climate change risks: a regional response from North Central Victoria, Australia. Mitigation and Adaptation Strategies for Global Change, 12(5), 685–712.

Kaplan, J., Frias, L., & Mcfall-Johnsen, M. (2020). A Third of the Global Population Is on Coronavirus Lockdown. Here Is Our Constantly Updated List of Countries and Restrictions. https://www.scirp.org/reference/referencespapers?referenceid=3732101

Kelly, P. M., & Adger, W. N. (2000). Theory and practice in assessing vulnerability to climate change and facilitating adaptation. *Climate Change*, 47, 325-352.

Kim, Y.H, Barber, N, Kim, D.K. Sustainability research in the Hotel Industry, Past, Present and Future, Journal of Hospitality Marketing and Management Vol 28, Issue 5, 2019

Lawrence, Martha, (2025) Driving green: the environmental impacts of EVs, https://www.edfenergy.com/energywise/electric-cars-and-environment.

Mazumder, S., & Chatterjee, A. (2007). Green Supply Chain as a competitive advantage enabler in the FMCG sector. Retrieved from http://www.bp-council.org/bpcpdf/Green Supply Chain as a competitive advantage enabler in the FMCG.pdf.

Mehta, A and Anjal Prakash, Hidden impact of semiconductor manufacturing on climate change, TOI https://timesofindia.indiatimes.com/blogs/voices/hidden-impact-of-semiconductor-manufacturing-on-climate-change/

McCarthy, J., Canziani, O. F., Leary, N. A., Dokken, D. J., & White, K. S. (2001). *Climate Change: Impacts, Adaptations and Vulnerability in Developing Economies,* World Bank, Washington, DC.

Mitchell & van Aalst, 2008, Tom Mitchell and Dr. Maarten van Aalst, A Review for DFID 31st October 2008. (UN water, Climate change is mainly about water.

Mitchell, T., & Tanner, T. (2006). *Adapting to climate change challenges and opportunities for the development community* (pp. 5-36). Institute of Development Studies, Tear Fund.

Mordor Intelligence (2021). Viscose staple fiber market - growth trends, COVID19 impact, and forecasts. Morlet, A., Opsomer, R., Herr.

Munasinghe, M. (2008). Addressing the Sustainable Development and Climate Change Challenges Together: Applying the Sustainomics Framework. Procedia Social and Behavioral Sciences, 41(2010), 6634-6640.

Palaksha, C., Nachiketas, N., & Rao, P. (2017). Industry initiatives to addressing water concerns in India today: The inward and the outward looking possibilities. *International Journal of Business Ethics in Developing Economies,* 6(1), 1-9.

Niinimäki 1. Greg Peters 2, Helena Dahlbo 3, Patsy Perry 4, Timo Rissanen 5 and Alison Gwilt 6 The environmental price of fast fashion Nature Reviews | Earth & Environment volume 1 | April 2020 | 189-199.

Patil, P.H. (2016), An Overview of Indian FMCG Sector, PARIPEX - INDIAN JOURNAL OF RESEARCH, Volume: 5 | Issue: 2 | February 2016.

Rao, Purba & Thamizhvanan, A (2014), "Impacts of climate change, Survey of mitigation and adaptation strategies of junior corporate executives in India", International Journal of Climate Change Strategies and Management. Vol.6 No.4.

Rao, P. (2020) "Green Supply Chain Management in Hotel Azure, a case in sustainability, Journal of Supply Chain Management Systems. 9(2&3):28-38.

Rao, P. (2019) "Green Supply Chain Management: A Study Based on SMEs in India," *Journal of Supply Chain Management Systems*. 8(1):15-24.

Rao, P. (2020) "Exploring Private Sector Initiatives for Adaptation to Climate Change Impacts on Water". *International Journal of Business Analytics and Intelligence* 8 (1), 53-63.

Rao, P. and A.R. Das, (2020), Corporate Managers and Climate Change Impacts, Great Lakes Herald, September 2020 Vol 14 Issue no 2.

Rao, P. (2002). Greening the supply chain: A new initiative in Southeast Asia. *International Journal of Operations and Production Management*, 22.

Rao, P., Pulupudi, R., & Sen, S. (2017). An application of structural equation modelling to determine the inclusion of climate change topics in MBA education. International Journal of Business Analytics and Intelligence, 5(1), 15-25.

Rao, P. (2001). Environmental Management Systems in Southeast Asia. Asian Institute of Management, Philippines.

Rao, Purba (2002), "Greening of the Supply Chain: a new initiative in South East Asia", International Journal of Operations and Production Management, 22(6), 632-655.

Rao, Purba. (2004). Greening the Supply Chain –A Guide for Managers in South East Asia. Published by Asian Institute of Management. Philippines.

Rao, Purba and Diane Holt, (2005), "Do Green Supply Chains lead to competitiveness and economic performance?" International Journal of Operations and Production Management, 25(9), 898-916.

Rani, G. and Hooda, K. (2013), Corporate Social Responsibility: Review of Literature, International Journal of Social Science & Interdisciplinary Research Vol. 2 (6), June (2013)

Sharma, T., Chen, J.S. Ramos, W. D. and Sharma, A. (2023) Visitors' eco-innovation adoption and green consumption behaviour: the case of green hotels, Indiana University Bloomington, Bloomington, Indiana, USA, and International Journal of Contemporary Hospitality Management © Emerald Publishing Limited, May 2023.

Seneviratne, S.I., et al. (2012) Changes in Climate Extremes and Their Impacts on the Natural Physical Environment. In: Managing the Risks of Extreme Events and Disasters to Advance Climate Change Adaptation, 109-230.

Singh, S.P., Sharma, N., Chandrakant, S.A. and Singh, S.P. (2021) "Electric Vehicles in India, A Literature Review", *7th International Conference on "New Frontier in Energy, Engineering and Science (NFEES),* 19-20 March 2021, ACEIT Jaipur.

Skoufias, E., Essama-Nssah, B., & Katayama, R. (2011). Too little, too late: Welfare impacts of rainfall shocks in rural Indonesia. Policy Research Working Paper 5615. The World Bank, Washington, D.C.

Sonia I. Seneviratne (Switzerland), Neville Nicholls (Australia) (2012), Changes in climate extremes and their impacts on the natural physical environment

(http://www.globalwaterforum.org/2016/01/04/strategies-for-countries-to-mitigate-the-risks-of-extreme-water-scarcity-and-dependency-in-a-sustainable-way/

Thakkar, H. (2010). India's tryst with the big irrigation projects. Retrieved from http://www.sandrp.in/ irrigation/Failure_of_ Big_ Irrigation_Projects_and_ Rainfed_Agriculture_0510.pdf

H. Turral, H., Burke, J., Faures, Jean Marc, (2011) Climate Change, Water and Food Security. https://www.researchgate.net/ publication/285797606_Climate_change_water_and_food_security

Triwadiantini, T. (2010). T. (2010). Contributor. Jakarta, Business, February 22 2010. Indonesia.

Verma, S, Malhotra, V, Rao, Purba (2017), Exploring Initiatives to Measure and Reduce Carbon Footprints Across Indian FMCG Supply Chain, Journal of Supply Chain Management System, Volume 6 Issue 3 July 2017

Vladimir, Kobayashi B., Stefan T. Mol Hannah A. Berkers, Gábor Kismihók, and Deanne N. Den Hartog, (2018) *Text Mining in Organizational Research, Organizational Research Methods* Volume 21, Issue 3, July 2018, Pages 733-765.

Wells, P. (2025) "India's transition to electric vehicles", https://www.polytechnique-insights.com/en/columns/economy/indias

Wilk, J., & Wittgren, H. B. (eds.) (2009). Adapting water management to climate change (Swedish Water House Policy Brief No. 7). Stockholm: Stockholm International Water Institute. WHWaterClimate.pdf

Youssef, A.B., Adelina Zeqiri2 Circular Economy and Sustainability,

https://doi.org/10.1007/s43615-021-00141-x Hospitality Industry 4.0 and Climate Change,

www.ingramcontent.com/pod-product-compliance
Lightning Source LLC
LaVergne TN
LVHW022216200926
848481LV00006B/1994